DEBT AND DEVELOPMENT IN SMALL ISLAND DEVELOPING STATES

Debt and Development in Small Island Developing States

Damien King and David F. Tennant

palgrave
macmillan

DEBT AND DEVELOPMENT IN SMALL ISLAND DEVELOPING STATES

First published in 2014 by
PALGRAVE MACMILLAN®
in the United States—a division of St. Martin's Press LLC,
175 Fifth Avenue, New York, NY 10010.

Where this book is distributed in the UK, Europe and the rest of the world, this is by Palgrave Macmillan, a division of Macmillan Publishers Limited, registered in England, company number 785998, of Houndmills, Basingstoke, Hampshire RG21 6XS.

Palgrave Macmillan is the global academic imprint of the above companies and has companies and representatives throughout the world.

ISBN: 978–1–137–39712–6

Library of Congress Cataloging-in-Publication Data

Debt and development in small island developing states /
[edited by] Damien King, David F. Tennant.
pages cm
Includes bibliographical references and index.
ISBN 978–1–137–39712–6 (hardback : alk. paper)
1. Debts, External—Developing countries. 2. States, Small—Economic conditions. 3. Islands—Economic conditions. 4. Economic development—Developing countries. I. King, Damien, 1955– II. Tennant, David F., 1975–

HJ8899.D4155 2014
336.3′4091724—dc23 2014026170

A catalogue record of the book is available from the British Library.

Design by Newgen Knowledge Works (P) Ltd., Chennai, India.

First edition: December 2014

10 9 8 7 6 5 4 3 2 1

Transferred to Digital Printing in 2015

To my heroes, three young men who inspire me and from whom I learn so much—Keiran, Darren, and Brandon

Damien

To Sandria, Sean, and Elia—my joy and inspiration—and my Father above—my light and foundation

David

Contents

Illustrations

Tables

Figures

Boxes

Acknowledgments

The idea for this book began with a project on Public Sector Debt in the Caribbean, commissioned by the Caribbean Development Bank (CDB), which we carried out under the auspices of the Caribbean Policy Research Institute (CaPRI). We are grateful to the CDB and to CaPRI for giving their permission for a few sections of that report to be reproduced in this volume. When we decided to expand the work done on that project to include an analysis of debt in all Small Island Developing States (SIDS), we relied heavily on the expertise of Michele Robinson—an experienced public debt manager, and now an independent consultant, who has provided technical assistance to countries in Africa, the Caribbean, the Pacific, and Eastern Europe. We are grateful for her guidance and contributions to this book. We must also express our appreciation to the other contributors to this collection, each of whom has enriched the output through their academic and/or policy expertise. Their adherence to the vision and guidelines provided, while infusing their own experiences and research, have allowed this collection to coalesce as a cohesive and comprehensive body of work.

Much coordination is required to pull together such a collection within stringent timelines and guidelines, and to this end, we are grateful to our Project Manager, Altricia Dawson, and our team of research assistants. Finally, we would not be able to complete this collection without the support of our employer, The University of the West Indies (Mona), and the helpful publishing team at Palgrave MacMillan.

Chapter 1

Debt and Development in SIDS: An Urgent Call for Action

David F. Tennant

The years 2008 to 2011 have marked a period of unprecedented global instability, recession, and crisis. Poverty and deprivation in developing countries have been exacerbated because of the increases in food and fuel prices. Also, because of the financial crisis–induced recession in the developed world, reduced demand for developing country exports, reduced private financial flows, and falling remittances have all adversely affected livelihoods in poor countries.[1] This has all been occurring while greenhouse gas emissions continue to increase, causing further rises in temperatures and sea levels.[2]

The prescribed medicine has often been bitter, even if necessary. Reductions in government subsidies and expenditures, particularly with regard to basic services, have disproportionately affected the poor and vulnerable.[3] Even the prospects of economic recovery are viewed as harbingers of worsened conditions for some, as recovery in global economic growth is likely to be accompanied by a resumption of food-price inflation, as well as further increases in emissions in the absence of determined action to shift to low-carbon economic models.[4]

To protect the world's poor and vulnerable from the continuing impacts of the Triple F crises (food–fuel–financial) and impending effects of a likely climate change crisis, developing countries need resources. Herein lies the often unacknowledged Fifth Crisis facing many poor countries—the burden of increasingly unsustainable debt, the servicing of which precludes growth-inducing and poverty-reducing government expenditures. The recent Triple F crises have significantly impacted public debt levels and trends.[5] Between 2007 and 2009, average debt levels increased by about 20 percent (in real terms) in countries that did not experience systemic financial crises and by about 75 percent in countries that did.[6] For countries with already high debt burdens, these increases were unsustainable and constituted a debt crisis.

This book examines the trend of high and rising debt levels in many Small Island Developing States (SIDS). It draws on the expertise of established researchers and public officials from SIDS to present a cohesive body of research examining the nature, accumulation, sources, causes, and impact of debt in SIDS. By so doing, it highlights the essential elements of an agenda for achieving debt sustainability in SIDS.

This introductory chapter contextualizes the analysis by providing answers to four key questions, which will justify the importance of this book. These are:

- Why Debt?
- Why SIDS?
- Why Now? and
- To What End?

Why Debt?

The 2015 deadline for the Millennium Development Goals (MDGs) is fast approaching and in some areas progress has been stubbornly limited. As an example, it is projected that poverty reduction efforts will not be sufficient to meet the MDG targets in Sub-Saharan Africa, Western Asia, and the Oceania Region. Northern Africa, Sub-Saharan Africa, South-Eastern Asia, Western Asia, the Oceania Region, Latin America, and the Caribbean all struggle to provide productive and decent employment for its citizens. Progress in promoting gender equality in employment and political representation remains limited across most regions, as has progress in improving access to reproductive health.[7]

By contrast, the United Nations (UN) reports significant progress on its eighth goal—developing a global partnership for development—with much success being noted in the efforts to deal comprehensively with developing countries' debt. In this regard, the UN (2013) notes that "debt service ratios are one-quarter less from their 2000 level, lessening the financial burden on developing countries. Better debt management, the expansion of trade and, for the poorest countries, substantial debt relief have reduced the burden of debt service" (1).

If the UN's assessment is correct, then the obvious question must be asked—why does this book focus on debt when there are so many other more pressing issues to be addressed? This is a good question, which necessitates a closer interrogation of the data.

First, note that although there has been progress in improving debt service ratios in numerous countries, the ratios in a number of countries remain high and unsustainable. Figure 1.1 shows that in 2012, 22 countries had debt service ratios of above 15 percent, the conservative guideline for what is sustainable. All 22 are developing countries, with 14 being classified as upper middle income, 7 as lower middle income, and 1 as low income (see figure 1.2). This income classification indicates that all but one of these countries have no chance of debt relief under the Heavily Indebted Poor Countries (HIPC) initiative.

The debt service ratio measures principal and interest repayments on external debt scaled to the country's exports of goods and services. Solely using this ratio to measure progress in dealing with developing countries' debt can be misleading. This is because numerous studies have shown that there has been a trend of high and increasing reliance on domestic debt in many countries, which is not captured by the debt service ratios.[8] Reinhart and Rogoff (2008) have noted that "for most countries..., domestic debt has been large and highly significant" (9).[9] This recognition is important, as it provides an explanation for why many countries default on (or restructure) their debt even when the debt service ratio

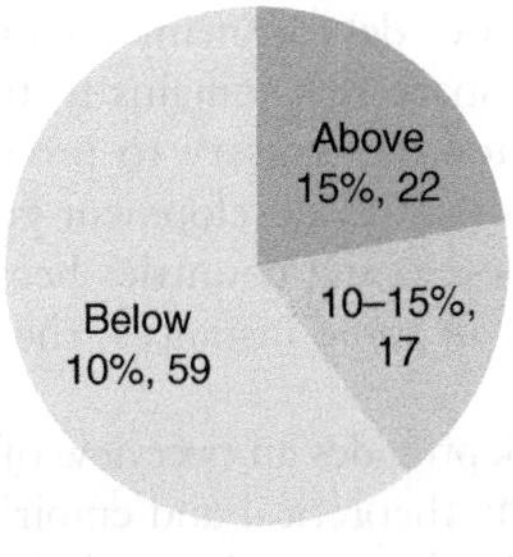

Figure 1.1 Debt service ratios by number of countries (2012)

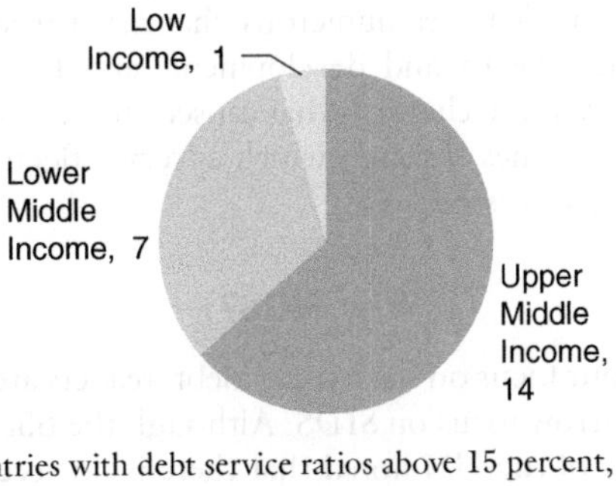

Figure 1.2 Countries with debt service ratios above 15 percent, by income (2012)

is seemingly low. When domestic debt obligations are accounted for, severe fiscal duress is often revealed. High debt burdens are thus likely to be much more pervasive than suggested by the UN MDG Progress Report.

In addition to this, the changing composition of debt in many countries has introduced varied types of risk, which, if not properly managed, can have adverse impacts. High shares of domestic debt can lead to crowding-out in private sector lending, precipitate weakening bank efficiency, and cause inflationary risks; while high shares of external debt expose the country to the risks of adverse movements in the exchange rate. Borrowing from bilateral or multilateral donors generally affords borrowers the most concessional terms, but loans are generally earmarked for projects or specific programs identified as part of a country strategy. By contrast, although sovereign borrowers have full discretion in the use of the funding obtained on international capital markets, borrowing on commercial terms is typically more costly. High shares of floating rate debt, and short-term debt expose the debt portfolio to the risk of adverse movements in interest rates or market conditions, which contribute to higher borrowing costs and an increased fiscal burden.

This then leads us to the crux of the matter—why focus on debt? Because the fiscal burden of debt in many countries remains high, and severely limits the fiscal space that remains to fund public investments and provide the social goods necessary to promote economic growth and development. The post-2015 development goals will continue to be unachievable for many developing countries because of the persistently high levels of debt, and, in some instances, the increased cost of such debt.

Chapter 2 of this book provides an overview of the debt–development dilemma by reviewing the theoretical and empirical literature on debt's impact on growth and development. It concludes that low to moderate levels of debt can be beneficial to development if it is properly used, but high levels of debt tend to be deleterious to targeted developmental outcomes. The chapter identifies numerous channels through which debt impacts economic growth and development, and highlights complex relationships, wherein the channels that caused the accumulation of debt often were the same ones through which adverse effects were felt when debt thresholds were crossed.

Why SIDS?

Having justified our focus on the issue of debt, readers are probably wondering why the narrow focus on SIDS. Although the 65th Session of the General Assembly of the UN noted that there is no accepted definition of a SIDS, it did highlight a number of widely accepted characteristics of the 38 UN member states that form part of this grouping.

> Small Island Developing States share very high levels of intrinsic vulnerabilities, especially to external shocks. The high levels of vulnerability of the natural, economic, and social systems of Small Island Developing States arise from the following characteristics: Small size; Remoteness; Vulnerability to external (demand and supply-side) shocks; Narrow resource base; and Exposure to global environmental challenges.[10]

We focus on SIDS in this book because they are among the most highly indebted countries in the world. When the public debt to gross domestic product (GDP) ratios of middle-income countries are compared as at the end of 2012, 12 of the 20 most indebted countries are SIDS. Of the five middle-income countries classified as having very high debt (with a debt to GDP ratio of above 90 percent), four were SIDS, with three of the four having debt to GDP ratios greater than 100 percent.

SIDS have on average, higher incomes than the landlocked developing countries.[11] This has important implications for debt, as only a very few SIDS have qualified for debt relief under the HIPC initiative. At the same time, however, the UN has acknowledged that:

> The recent development progress made by Small Island Developing States might be jeopardized by major ongoing shocks. Their vulnerability has increased due to climate change and was most recently demonstrated by the global financial crisis of 2007–2010, the food and fuel crises of 2007–2008 and the large-scale natural disasters, which occurred in 2009–2010. For example, the food crisis has had a severe impact on the poor in the Small Island Developing States, most of which are net food importers.[12]

The unique vulnerability of SIDS thus places a heavy strain on their governments to attempt to address the environmental, social, and economic challenges. Their ability to implement proactive policies and to respond to natural and external shocks is, however, for some SIDS, severely curtailed by the fiscal burden of the debt. The post-2015 development agenda of the SIDS thus correctly places their debt burden in the spotlight.[13]

Chapters 3, 4, and 5 of this book provide detailed analyses of the debt experiences of SIDS in the Caribbean, Pacific, and the AIMS regions.[14] In each chapter an introduction to the region is followed by an overview of public debt. The analysis of the debt profile of the countries includes an examination of the evolution and composition of the debt. A discussion of the factors currently affecting debt trends in the respective regions, as well as those expected to affect trends in the near future, follows. The analysis provided in these chapters highlight the varied experiences that SIDS have had with debt accumulation and debt management in their recent past.

Caribbean SIDS were shown to be most affected by high and rising public debt, with accommodative fiscal policies, natural disasters, and economic shocks contributing to the accumulation. SIDS in the Pacific, by contrast, are shown to have relatively low debt levels, which has been at least partially attributed to their typically low income levels, which allow them to secure loans on highly concessional terms. The AIMS grouping of SIDS includes countries with quite different economic and sociopolitical histories, resulting in widely varying debt experiences. Common, however, with the Caribbean is the fact that debt accumulation in AIMS SIDS has been largely influenced by international shocks and natural disasters. Additionally some countries have been adversely affected by political instability, while others have benefitted tremendously from debt relief. The depth of analysis in these chapters, combined with the richness of the varied narratives provide much scope for drawing lasting lessons on debt accumulation and management from and for SIDS.

Why Now?

The high debt burdens faced by many SIDS, added to the economic, social, and environmental vulnerabilities that they face, combined with the pressing need but limited ability of their governments to implement proactive and adaptive policies, strongly indicate that urgent action is needed to create fiscal space. For a number of SIDS this will require reduced debt accumulation, along with improved debt management.

Chapter 6 strongly makes this case for the most heavily indebted Caribbean SIDS by projecting the likely outcomes of debt to GDP ratios up to 2020, based on the trajectories of domestic and foreign currency debt. It examines the policy measures currently being implemented in each of the countries and assesses whether they are adequate to place the countries on a sustainable debt trajectory. The analysis indicates that the majority of the highly indebted SIDS should not be satisfied with the measures currently being implemented, as they will not by 2020 result in sustainable levels of debt. Furthermore, even in the few countries that are implementing sufficiently stringent policies to curtail increasing debt levels, macroeconomic shocks have been shown to easily and significantly throw them off track. At the other extreme, the results in chapter 6 also clearly show that if the governments of the most indebted countries choose to do nothing about the debt levels, debt defaults are highly likely.

This places the governments of these SIDS in a tremendously difficult situation, as for most countries, whereas the option of doing nothing about debt will lead to default, current efforts are inadequate and highly vulnerable to shocks. This suggests that closer attention has to be placed on the debt reduction and debt management efforts. A critical part of sustainable debt reduction is curtailing present and future accumulation. Effectively doing so requires a thorough understanding of the sources of debt accumulation.

Chapter 7 identifies the sources of debt accumulation. It conducts a rigorous debt decomposition exercise for the highly indebted SIDS in the periods wherein there was sustained debt accumulation. The analysis highlighted two key drivers of debt accumulation in SIDS: (1) fiscal imbalances, often due to infrastructural reconstruction following a natural disaster or to developmental capital expenditure; and (2) contingent liabilities, the sources of which include weak contingent risk-management, public enterprise holdings, weaknesses in public sector management, and inadequate financial sector management.

Many of the issues highlighted in chapter 7, point to the institutional underpinnings of debt accumulation and debt management, all of which must be urgently addressed if sustainable debt levels are to be realized in the highly indebted SIDS. Chapter 8 delves deeper into these issues by exploring the institutional foundations of the debt in SIDS, and the institutional challenges for debt management in these countries. It highlights the institutional requirements for sound public debt management, and then examines the unique challenges of island-state economic governance, which constrain SIDS in the effective implementation of the debt management function.

Why focus on debt in SIDS now? Because many SIDS face severe debt burdens that are projected to worsen if policy changes are not implemented. Fundamental changes in debt management and public sector management are urgently required if meaningful change is to be realized before economic and developmental outcomes are further stymied.

To What End?

What do we hope to achieve by compiling these analyses on various aspects of debt in SIDS? Simply put, we want to make a difference. Although this book is completely based on the principles of academic rigor and objectivity, it is not solely an academic exercise. We hope that the results of the analyses can be used to guide policy makers in SIDS and those within the global financial architecture as to the most effective ways of realizing debt sustainability in SIDS.

Chapter 9 thus tackles the controversial issue of debt default, by assessing recent actions taken by selected SIDS to reduce their exposure to debt through different types of debt restructuring exercises. It examines the current debt restructuring mechanisms available to SIDS and the extent to which using these mechanisms have provided debt relief. The chapter concludes that the debt restructuring exercises utilized by most SIDS have provided only temporary cash relief and little debt reduction. This is because concerns with loss of access to international capital markets and, in the case of domestic debt restructurings, concerns about financial market stability, constrained many restructuring exercises to focus on reducing interest payments and lengthening tenors, rather than debt cancelations (i.e. to focus on liquidity rather than solvency). For the few SIDS that received a haircut, lost market access was temporary, and debt moved closer to sustainable levels. The chapter thus concludes that for some SIDS, haircuts may be the only viable option for debt sustainability.

This is critical, because debt sustainability impacts the achievement of all developmental outcomes. Chapter 10 explores in some depth the link between debt sustainability and the goals of sustainable development in SIDS. It shows how high levels of public debt in SIDS can adversely affect each of the three pillars of sustainable development—economic development, social development, and environmental protection. The unique challenges faced by SIDS in achieving each of these goals are highlighted, with the exacerbating effect of high debt burdens being emphasized. The relationship is, however, complex, with the achievement of debt sustainability being in turn impacted by the extent to which the goals of sustainable development are achieved, particularly that of economic growth. The chapter thus advocates for the prioritization of debt sustainability in the development agenda for SIDS.

Chapter 11 provides the information needed for such advocacy by presenting the essential elements of an agenda for debt sustainability in SIDS. It focuses on the institutional foundations for fiscal outcomes and debt accumulation, and provides recommendations for improved fiscal management, debt management, and liability management in SIDS.

Chapter 12 concludes the book. The synthesis of the analyses and results of the previous chapters is used to emphasize the urgent call for action for governments and global partners to implement the necessary measures to address the high debt burdens of many SIDS, as they seek to achieve sustainable development in spite of their unique vulnerabilities.

Notes

1. Heltberg, Hossain, and Reva (2012).
2. Addison, Arndt, and Tarp (2010).

3. See, for example, http://www.odi.org.uk/publications/6049-food-finance-fuel-crisis-women-children-nigeria.
4. Addison, Arndt, and Tarp (2010).
5. Caner, Grennes, and Koehler-Geib (2010).
6. Reinhart and Rogoff (2010).
7. http://www.un.org/millenniumgoals/pdf/report-2013/2013_progress_english.pdf.
8. See, for example, Hanson (2007) and Reinhart and Rogoff (2008).
9. Reinhart and Rogoff (2008) further note that "nothing about the maturity structure or interest rates paid on these debts lends justification to the common practice of ignoring them in calculations of debt sustainability..." (9).
10. http://www.un.org/en/ga/65/meetings/mauritiusreview.shtml.
11. http://www.un.org/en/ga/65/meetings/mauritiusreview.shtml
12. http://www.un.org/en/ga/65/meetings/mauritiusreview.shtml.
13. http://www.sids2014.org/index.php?menu=1537.
14. AIMS includes the SIDS in the Atlantic and Indian Oceans, as well as the Mediterranean and South China Seas.

References

Addison, Tony, Channing Arndt, and Finn Tarp. 2010. "The Triple Crisis and the Global Aid Architecture." *UNU-WIDER Working Paper No. 2010/01.*

Caner, Mehmet, Thomas Grennes, and Fritzi Koehler-Geib. 2010. "Finding the Tipping Point—When Sovereign Debt Turns Bad." *World Bank Policy Research Working Paper 5391.*

Hanson, James A. 2007. "The growth in government domestic debt: changing burdens and risks." *Policy Research Working Paper Series 4348.* The World Bank.

Heltberg, Rasmus, Naomi Hossain, and Anna Reva. 2012. *Living through Crises: How the Food, Fuel, and Financial Shocks Affect the Poor.* Washington, DC: World Bank.

Reinhart, Carmen M. and Kenneth S. Rogoff. 2008. "The Forgotten History of Domestic Debt." *NBER Working Paper 13946.*

———. 2010. "The Forgotten History of Domestic Debt." http://scholar.harvard.edu/files/rogoff/files/forgotten_history_of_domestic_debt.pdf.

United Nations (UN). 2013. *We Can End Poverty: MDGs and Beyond 2015—Fact Sheet, Goal 8.* New York: United Nations. http://www.un.org/millenniumgoals/pdf/Goal_8_fs.pdf.

World Bank. 2013. *World Development Indicator Database.* Washington, DC: World Bank. http://data.worldbank.org/indicator/NY.GNP.PCAP.CD.

Chapter 2

The Debt-Development Dilemma: Challenges, Channels, and Complexities*

David F. Tennant

The recent global financial and economic crisis has significantly impacted public debt levels and trends.[1] Between 2007 and 2009, average debt levels increased by about 20 percent (in real terms) in countries that did not experience systemic financial crises and by about 75 percent in countries that did (Reinhart and Rogoff 2010). This is attributed to direct bailout costs in some countries, the adoption in many countries of stimulus packages to deal with the recession, and substantial declines in government revenues that affected most countries.

Before the crisis, however, during the 1990s and early to mid-2000s, the debt ratios of developing countries were generally improving. The average debt to GDP ratio of developing countries stabilized in the first half of the 1990s and then started to decline rapidly in the new millennium, up till about 2007.[2] It is important to note, though, that this trend masked an important cross-country heterogeneity. Small countries have debt ratios which are substantially higher than those of larger economies, and this difference has been increasing over time (Panizza, Sturzenegger, and Zettelmeyer 2010).

At the end of 2012, a half of the 20 most highly indebted countries were from the narrow grouping of countries classified as Small Island Developing States (SIDS). Six of these countries were among the top ten most highly indebted. Subsequent chapters will explore the debt experience in these countries and highlight reasons for the accumulation of the debt. This chapter examines the developmental impact of debt accumulation. It begins by highlighting the non-linear relationship between debt and development by summarizing the main theoretical and empirical studies in this area, and then explains this relationship by examining a number of specific channels through which debt impacts economic growth and development.

Debt and Development: Theory and Evidence

It is widely accepted that most developing countries cannot grow without borrowing to finance the technology gains and capital deepening that precipitate economic progress. High levels of debt are, however, increasingly asserted to have deleterious effects on economic development.[3] Much of the literature in this area focuses on the relationship between debt and economic growth. More recently, though, there have been a number of articles that have examined the ways in which debt can impact poverty levels and alleviation efforts. This section briefly summarizes theoretical and empirical literature on the impact of debt on a country's developmental fortunes.

Debt and Economic Growth—Theory and Intuition

Several authors have highlighted the positive impact that public debt can have on the effective functioning of an economy. At a very basic level, public debt enables fiscal authorities to play their role in stabilizing the macroeconomy, which is a critical ingredient for growth creation. As an example, Bourne (2010) notes that countering the effects of external economic shocks and natural hazard events are legitimate motivations for debt accumulation. When governments borrow, they are able to smooth taxes in the face of variable expenditures. "Since part of the tax rise needed to fund higher current consumption is postponed, public debt may rise, at least up to a point, without growth necessarily slowing."[4] This is because debt helps to "smooth consumption not only through the lifetime of individuals who are currently alive, but also across generations. To the extent that future generations will be richer than the current ones—because they will have a combination of more human capital and more productive technology—a transfer from future to current generations can raise society's intertemporal welfare" (Cecchetti, Mohanty, and Zampolli 2011, 3).

More fundamentally, numerous theoretical works suggest that low to moderate debt levels are positively associated with economic growth. At such debt levels, a higher debt/GDP ratio loosens the credit constraints, makes more resources available for investments, and fosters economic growth (Caner, Grenner, and Koehler-Geib 2010). This is captured in endogenous growth models through capital stock augmentation, which includes enhanced human capital through public investment in education and training, and the effect of technical change embodied in new capital goods.[5] The underlying assumption is that the borrowed funds are used to finance productive investments (Schclarek and Ramon-Ballester 2005). Where this is the case, authors such as Aschauer (1989) and Khan and Kumar (1997) show that public investment expenditures have positive and significant effects on economic growth.

The theorized positive linear relationship between debt and economic growth, has, however, recently been increasingly rejected in favor of a virtual consensus around a nonlinear relationship between debt and growth. In this theorized relationship, the growth-inducing benefits of low to moderate levels of debt are eventually reversed as debt accumulates beyond a certain threshold. This threshold is theorized to occur when there is a debt overhang, which is defined as "the presence of an existing inherited debt sufficiently large that creditors do not expect with confidence to be fully repaid" (Krugman 1988).[6] Where this occurs, it is argued that the heavy debt burden reduces a debtor country's incentive to invest. This is because when the country is unable to service its debt in full, the actual payments will depend on its economic performance. In this circumstance, "if output increases as a result of higher investment, some of the additional proceeds accrue to creditors in the form of higher debt service payments. In this way, debt acts as a marginal tax on investment" (Husain 1997, 519).

Deshpande (1997) thus notes that this scenario is the basis for the now widely touted debt overhang hypothesis (DOH), according to which "investment is discouraged if the debt burden is so large that the debtor country is unable to meet its payment obligations in a normal way, and involuntary lending takes place. Thus the DOH describes a situation in which the foreign debt is not merely large, but one in which the existence of foreign debt distorts the relevant margins considered for production and investment decisions" (171). Clements, Bhattacharya, and Nguyen (2003) explain by noting that where there is a debt overhang, "some of the returns from investing in the domestic economy are effectively taxed away by existing foreign creditors, and investment by domestic and foreign investors—and thus economic growth—is discouraged" (3–4). In other words, "potential investors will fear that the more is produced, the more will be taxed by creditors to service the external debt, and thus they will be less willing to incur investment costs today for the sake of increased output in the future" (Pattillo, Poirson, and Ricci 2004, 5).[7]

Debt and Economic Growth—Empirical Evidence

Table 2.1 presents a summary of 19 empirical studies on the relationship between debt and economic growth. Immediately evident is the fact that the theorized positive linear relationship between public debt and economic growth has not received empirical support. None of the studies reviewed provided any evidence supporting such a relationship; instead, in the majority of the empirical work, the theorized nonlinear

Table 2.1 Tabular summary of the empirical evidence on the relationship between debt and economic growth

Author(s)	Primary Debt Indicator Used	Data	Methodology	Results				
				Linear Relationship	Non-Linear Relationship	Inconclusive/ No relationship	Debt Overhang Thresh'ld (percent of GDP)	Channel
Fosu (1996)[‘]	Ext Debt	SSA countries (1970–86)	OLS	-ve				
Clements et al. (2003)	Ext Debt	55 LICs (1970–99)	FE & SGMM		X		50 percent	Efficiency of resource use
Pattillo et al. (2004)	Ext Debt	61 LDCs (1969–98)	OLS, SGMM, diff-GMM		X			Physical capital accumulation and total factor productivity growth
Schclarek and Ramon-Ballester (2005)	Ext Debt	20 LAC countries (1970–02)	Dynamic system GMM	-ve				Capital accumulation growth
Shabbir	Ext Debt	24 LDCs (1976–03)	Linear panel data model	-ve				Crowding out
Cordella et al. (2005)	Nominal Debt & a proxy for NPV of Ext Debt	79 LDCs (1970–02)	OLS, SGMM		X		15–30 percent[ι]	Quality and volatility of investment and policy incentives
Elbadawi et al. (1997)	Nominal debt	99 LDCs	Quadratic equations		X		100 percent	
Imbs and Ranciere (2005)	Ext debt	87 LDCs (1969–02)	OLS, FE & SGMM		X		60 percent	Quantity of investments and conduct of government policy
Kumar and Woo (2010)	Public Debt	38 DCs and emerging economies[’] (1970–07)	OLS, FE & SGMM		X		90 percent	Labour productivity growth and quantity of investment
Reinhart and Rogoff (2010)	Public Debt[ʿ]	44 DCs & LDCs (spanning 2 centuries)	Correlations		X		90 percent[’]	

Caner et al. (2010)	Gross public debt	101 DCs & LDCs (1980–08)	Threshold LS, Pooled LS		X		77 percent (full sample); 64 percent (LDCs)	
Schclarek (2004)[i]	Gross government debt	24 DCs (1970–02)				X		
Checherita and Rother (2010)	Gross government debt	12 Euro Area countries (1970–10)	Panel fixed effects		X		90–100 percent	Private saving, public investment, total factor productivity, sovereign LT interest rates
Cecchetti et al. (2011)	Total non-fin debt	18 DCs (1980–10)	Growth regression		X		85 percent[g]	
Abbas and Christensen (2007)	Public Dom Debt	93 LICs and emerging markets (1975–04)	Granger causality regression		X		35 percent of bank deposits	Monetary Policy, Fin market development, Crowding out
Singh (1999)[i]	Dom Debt	India (1959–95)	Co-integration			X		
Maana et al. (2008)[j]	Dom Debt	Kenya (1996–07)	OLS	+ve				
Sheikh et al. (2010)	Dom Debt	Pakistan (1972–09)	OLS	+ve for debt stock; -ve for debt service				
Adofu and Abula (2010)	Dom Debt	Nigeria (1986–05)	OLS	-ve				

Notes: a: As quoted in Qureshi and Ali (2010).
b: This is for countries with good policies and institutions. Countries with bad policies and institutions have lower thresholds. Cordella et al (2005) also specify a debt irrelevance threshold of 70–80 percent of GDP.
c: With populations greater than 5 million.
d: Defined as gross central government debt, and includes domestic and external public debts.
e: This figure represents total public debt/GDP. For emerging markets there is a more stringent external debt/GDP ratio of 60 percent.
f: As quoted in Sheikh et al. (2010).
g: This figure is for government debt and household debt. The corporate debt threshold was computed at 90 percent of GDP.
h: As quoted in Sheikh et al. (2010).
i: As quoted in Sheikh et al. (2010).
j: The result was positive but insignificant.

relationship between debt and growth has been explicitly tested and confirmed. Eleven reputable empirical studies conducted since the early 2000s, have found a robust nonlinear relationship between debt and growth. They confirm the theory that reasonable levels of borrowing by a developing country are likely to enhance its economic growth, but that large levels of accumulated debt lead to lower growth.

The wide acceptance of this nonlinear relationship has naturally led to much discussion about the levels of debt that are considered to be "reasonable" versus "large." Caner, Grenner, and Koehler-Geib (2010) note that the increase in public debt associated with the current global economic crisis "has raised concerns over whether it is starting to hit levels at which it might slow economic growth" (2). They raise questions as to how strong the growth impact would be if debt surpassed the threshold, and "what would happen if debt stayed at elevated levels for an extended period of time" (2). While these questions are now increasingly relevant for a larger number of countries, they have been quite relevant to numerous SIDS for many years.

The penultimate column in table 2.1 highlights the results of studies that have attempted to compute the debt overhang threshold as a percentage of GDP. Whereas they may not all be directly comparable because of different debt measures used, the wide variation in estimates is nonetheless instructive, with figures ranging from 15 percent to 100 percent. Caner, Grenner, and Koehler-Geib (2010) explain by noting that there are theoretical and empirical reasons why debt threshold levels may vary across countries, particularly by country income. "Debt may play out differently in low-income countries, because of less developed domestic financial markets, a different degree of openness, and different institutions" (2). Sahay (2006), however, notes that "in general, public debt/GDP ratios over 50 percent to 60 percent are considered high" (7). By that measure, it is clear that numerous small states are highly indebted, with the average debt to GDP ratio for small states being 60.5 percent in 2012.[8] The average ratio for Caribbean SIDS is particularly high at 87.9 percent in 2012.[9] By these estimates, numerous small states, especially those in the Caribbean region, have surpassed the debt threshold levels computed in the majority of studies collated in table 2.1.

This highlights the importance of determining the effect on growth if debt stays at elevated levels for an extended period of time. Caner, Grenner, and Koehler-Geib (2010) estimated forgone growth as a result of exceeding the debt threshold for 29 countries. The annual percentage point loss in real GDP growth ranged from a negligible amount for Greece and the Philippines to 4.7 percent for Nicaragua. This translated to a cumulated loss over 28 years of 0.5 percent and 1.2 percent for Greece and the Philippines, respectively, and 264.6 percent for Nicaragua. Jamaica was the only SIDS included in that study, and her annual percentage point loss in real GDP growth was estimated as 0.2 percent, which cumulated to a 5.1 percent percentage point loss over 28 years. Because this is a relatively new area of study, these results have not yet been substantiated or refuted. They are nonetheless instructive as to the possible range of negative effects that could be incurred if debt thresholds are exceeded.

Debt and Poverty—Intuition and Evidence

Most of the research on the impact of debt has focused on its effect on economic growth. It is, however, now widely accepted that economic development encompasses the broader objectives of alleviating poverty and reducing inequitable income distribution. Whereas it is difficult (if not impossible) to sustainably achieve these goals in the absence of economic growth, such growth by itself does not guarantee their achievement.[10] Also important is the fact that the ultimate impact of any policy or shock on development can be felt through its effect on economic growth, and/or through a direct effect on poverty or distributional outcomes.

While the effect of debt on growth is well researched, very little has been written on its direct impact on, for example, poverty alleviation efforts. It is theorized that at high levels of debt, increasing debt service requirements could force governments to allocate resources away from developmental initiatives or welfare programs. Any reduction in public provision of health care and education clearly disproportionately impacts the poor, and highlights one possible avenue through which high levels of debt servicing can adversely impact development (independent of its impact on economic growth).

Although studies such as those by Kemal (2001) and Uzochukwu (2003) found that debt accumulation and servicing have adverse effects on the poor,[11] clear evidence as to the channels through which those impacts are felt was not derived. This is because whereas some studies found that debt service had a negative effect on total government spending, very few such studies disaggregated public expenditure. Fosu (2007) sought to fill this gap in the literature by examining the effect of increased debt servicing on fiscal allocations in 35 Sub-Saharan African economies. The results suggested a strong direct impact on developmental initiatives, as it was found that "a debt-servicing constraint would shift public expenditure away from the social sectors of health and education, and possibly from public investment" (16).

Empirical correlations of the relationship between debt and poverty are likely to understate the impact of indebtedness on needy households. Some of the socio-economic effects of a debt burden are manifest only when the government has to set about reducing it. Whenever governments choose to inflate their way out of domestic debt, rising inflation redistributes income from the poor to the rich and so is associated with rising poverty. It is also often asserted that in instances of forced fiscal contractions, social programs (e.g. poverty alleviation projects) are often substantially reduced because the poor tend not to have the organized clout to mount any effective opposition.

Rising public debt, then, is likely to be a threat to the standard of living of the lower socioeconomic groups especially. So whatever its effect on economic growth, the impact of debt on the other dimensions of development should be a concern for SIDS facing a growing debt problem.

Channels through which Debt Impacts Growth and Development

Whether debt's impact on development is felt through the effect on economic growth, or via its impact on poverty alleviation efforts, the intuition and empirical evidence is clear—low levels of debt can be beneficial to developmental outcomes provided that the debt is properly used, but high levels of debt tend to be deleterious to such outcomes. Not as clear, however, are the nuances of the relationship. This section examines in more detail the various channels through which these outcomes are manifested.

Debt and the Fiscal Balance

In principle, fiscal deficits can be financed through debt accumulation or money creation. Debt has tended to be the preferred option in many SIDS of late. Printing money has become less tolerable with the increased appreciation for the long run economic costs of inflation. Moreover, a number of SIDS are part of a currency union and so do not have the required authority over the central bank. Thus, increasing fiscal deficits have precipitated a massive buildup of debt in many countries.[12]

This accumulation of debt has, in turn, had implications for growth and development through both sides of the fiscal balance. From as early as 1965, Diamond theorized that external and domestic debt adversely impact the capital stock through the effect of the taxes needed to service the debt. This is because such taxes "reduce the available lifetime consumption of taxpayers, as well as their saving, and thus the capital stock."[13] High levels of debt also impact governments' spending decisions. It was previously noted that the large percentages of GDP required to service public debt have forced many governments to reduce expenditures, and have crowded-out pro-poor and growth enhancing spending (Abbas and Christensen 2007). The adverse growth impact of a reduction in spending on education must, however, be emphasized, as new growth theorists have highlighted the importance of quality education to human capital accumulation and innovation.[14] Cuts in public sector investment can also be growth-retarding, as reductions, particularly in physical infrastructure, tend to crowd out private investment, since they are often complementary to private sector activities.[15]

If debt management in a country is weak, the relationship between the fiscal balance and public debt is likely to result in worsened growth and development outcomes. The buildup in debt caused by persistent fiscal deficits will ultimately precipitate distortionary and/or austerity measures by governments as debt servicing requirements become burdensome.[16] Where debt is unsustainable, the requisite adjustment measures have to be carefully considered so as to minimize the possible adverse impacts on growth and development.

Debt and the Financial Sector

The spate of banking crises in the 1990s contributed to the rapid accumulation of public debt in many developing countries, as domestic debt was issued to recapitalize insolvent banks and to fund governments when the supply of foreign loans declined.[17] Since then "governments in both crisis and non-crisis countries have continued to rely increasingly on domestic debt" (Hanson 2007, 1). The recent global financial crisis has put further strain on debt in several developed and developing countries.[18] These episodes of crisis highlight the dialectical relationship between a country's financial sector and its level of domestic debt. In countries that have experienced a financial crisis, the issuance of domestic debt fosters the short-term stability of the financial system when the funds are used to resuscitate the failing institutions and restore confidence. Even in non-crisis countries, issuance of domestic debt can have positive spillovers for the financial sector, as it fosters the development of private capital markets and boosts private savings.[19]

Careful consideration of the impact of the changing composition of debt must, however, be emphasized, as the growth in domestic debt implies changing burdens and risks. The most severe negative effects of domestic debt are also channeled through the financial sector. The crowding-out effect of domestic debt on private investment is a serious concern. Bank credit to the private sector has been empirically proven to be a contributor to economic growth.[20] But when governments borrow domestically, they "use up domestic private savings that would otherwise have been available for private sector lending" (Abbas and Christensen 2007, 4).[21] As increasing public financing needs push up sovereign debt yields, this further causes a net flow of funds out of the private sector into the public sector, and pushes up private interest rates.[22] "In shallow financial markets, especially where firms have limited access to international finance, domestic debt issuance can lead to both swift and severe crowding out of private lending."[23] In most developing countries, only large, well-established firms have access to international finance, suggesting that the burden of crowding-out falls heavily on small and medium-sized enterprises and rural borrowers (Hanson 2007).

Empirical support for the crowding-out argument has been provided by Abbas and Christensen (2007). In a cross-country study of 93 developing countries, they found a nonlinear relationship between economic growth and public domestic debt measured as a ratio of deposits. Domestic debt was seen to support growth up to a ratio of 35 percent of deposits, but strangled it at higher levels of this ratio. Abbas and Christensen (2007), however, assert that "a higher level of domestic debt can be likely sustained without compromising growth if domestic debt is issued in the form of marketable securities, bears positive real interest rates, and is issued to investors outside the banking system" (21).[24]

Notwithstanding this, the higher interest rates that typically accompany reliance on domestic debt also affect financial institutions by creating an adverse selection problem. As interest rates rise, more conservative, risk-averse borrowers shy away from the credit market. A larger proportion of the persons applying for loans are thus those who are willing to take risky bets. The likelihood of default increases and so therefore does the banks' proportion of nonperforming loans.

The stability of the financial system is further threatened by the heightened risk of governments defaulting on domestic debt, as the debt servicing burden becomes overwhelming.[25] Because so many financial institutions are heavily invested in government bonds, any such default could impact their solvency. Hanson (2007) further argues that the risk for such banks not only arises in the extreme cases of default, but also even as government debt managers seek to ease their own risks by extending the maturity of their debt instruments and selling more debt with fixed interest rates. "This policy increases banks' liquidity risk systematically to the extent that the changes in government domestic debt create a mismatch with bank deposits. Interest rate risk increases to the extent that the domestic debt instruments increasingly carry fixed rates and are of longer maturity than deposits...In the case of a major rise in interest rates, these risks from debt management policies translate into actual bank liquidity problems, declines in banks' net income, and reductions in bank capital" (26).

Abbas and Christensen (2007) thus observe that "issuing domestic debt, whether to finance the fiscal deficit or to mop up monetary liquidity, involves a complex evaluation of the costs and benefits to the economy" (4). The link between financial sector stability and fiscal sustainability, through the level of domestic debt, is thus an important issue for SIDS, which has to be closely monitored.

Debt and Macroeconomic Policies and Institutions

The abovementioned nexus between fiscal and debt sustainability and financial stability can also impact a country's fundamental macroeconomic indicators. For example, burdensome domestic debt servicing requirements increase the incentive to monetize deficits, thus reducing the real burden of domestic debt in the resultant inflationary environment.[26] Such inflationary conditions can foster the creation of asset price bubbles, the bursting of which can lead to financial sector instability.[27]

On the other hand, where highly indebted countries have weak fiscal and financial systems, high fiscal deficits along with very limited possibilities for domestic borrowing, have led to overreliance on foreign debt. In the 1990s, "such countries were particularly subject to sudden stops in capital inflows that led to large devaluations, major financial disruptions, and output declines" (Hanson 2007, 22). In countries which rely on imports for a large proportion of capital goods, large and sudden exchange rate devaluations can have severe adverse effects on investments.[28]

It is not only possible for high levels of public debt to precipitate macroeconomic and financial instability, but it can also limit governments' ability to adequately respond to such instances of instability. Cecchetti, Mohanty, and Zampolli (2011) explain by noting that "when a crisis strikes, the ability of the government to intervene depends on the amount of debt that it has already accumulated as well as what its creditors perceive to be its fiscal capacity—that is, the capacity to raise tax revenues to service and repay the debt. Fiscal authorities may become constrained both in their attempt to engage traditional countercyclical stabilization policies and in their role as lender of last resort during a financial crisis" (4).

It is thus argued that high levels of public debt can limit critical government functions. Deshpande (1997) gives the example that "in a situation of debt overhang, with low rates of growth and possibly declining standards of living, a government would find it difficult to shift resources from consumption to investment as the people of the debtor nations would be inclined to believe that this shift would be utilized only to service debts, thus robbing them of potential benefits" (172–173).[29] Governments may therefore be "less willing to undertake difficult and costly policy reforms if it is perceived that the future benefits in terms of higher output will accrue partly to foreign creditors" (Pattillo, Poirson, and Ricci 2004, 3).[30] Politically costly but necessary policy reform efforts may therefore be postponed. For example, reliance on domestic debt may delay necessary tax mobilization efforts.[31]

Cordella, Ricci, and Ruiz-Arranz (2005) have shown that countries with better institutions and policies have higher debt thresholds. This enables them to benefit from the growth-enhancing effects of borrowing at higher levels of debt. By contrast, countries with weak policies and institutions are likely to have poor debt management mechanisms and thus suffer from more rapid debt accumulation. Once the debt threshold is crossed, the debt overhang hypothesis suggests that governments are no longer incentivized to undertake the costly institutional and policy reforms that would ensure a return to fiscal and debt sustainability.

Countries can thus get trapped in a high debt, low growth equilibrium. The experience in SIDS must be examined to ascertain whether such countries are either already in or are approaching such a situation. The policy and institutional framework must also be investigated to determine whether policies and institutions contribute to favorable outcomes, or whether they concretize the debt trap.

Debt and Private Investors

Many of the aforementioned consequences of debt impact economic growth through their effect on private investors and entrepreneurs. For example, the disincentive effect of the debt overhang led to capital flight from many severely indebted countries in the 1980s debt crisis, as, due to the fear of appropriation of their funds for debt servicing, private investors preferred to send their money out of the country[32]; crowding-out precluded private investors from accessing low-cost domestic funds following many of the financial crises of the 1990s; and, where governments postponed necessary policy reforms, the poorer policy environment affected the efficiency of investment and productivity of private producers in a number of developing countries.[33]

More subtly, however, debt overhang also depresses investment and growth by increasing uncertainty. Clements, Bhattacharya, and Nguyen (2003) explain by noting that "as the size of the public debt increases, there is growing uncertainty about actions and policies that the government will resort to in order to meet its debt servicing obligations . . . The extensive literature on uncertainty and investment suggests that in these circumstances, potential foreign investors will . . . exercise their option of waiting" (4). This occurs even if the economy's fundamentals are good and/or are improving, because the risk of default, fears of devaluation, and a cyclical borrowing and rescheduling cycle increase the volatility of private capital flows.

Domestic entrepreneurs will adjust to the uncertainty by reducing their planning horizon. Any investment that takes place in an uncertain environment is likely to be allocated to "activities with quick returns, rather than long-term, higher-risk irreversible investment which would be more conducive to long-run productivity growth. Misallocated resources and less efficient investment projects could thus contribute to slower productivity growth" (Pattillo, Poirson, and Ricci 2004, 3). The high levels of uncertainty related to the debt overhang are thus likely to reduce private investors' incentives to improve technology or to use resources efficiently.

Private investors will take more drastic action than opting to wait or shortening their planning horizons when the uncertainty increases beyond a certain point. Clements, Bhattacharya, and Nguyen (2003) note that "rapid accumulation of debt can also be accompanied by increasing capital flight if the private sector fears imminent devaluation and/or increases in taxes to service the debt" (4). Large public debts can also hinder further capital inflows and reduce indebted countries' access to international capital markets.

High levels of debt thus affect the productivity and profitability of private investment. This has clear deleterious effects on economic growth and development. It also has implications for debt sustainability. This is because the ability of the public sector to sustain a given level of debt depends on its ability to raise tax revenue from the private sector (Cecchetti, Mohanty, and Zampolli 2011). If the productivity and level of private economic activity diminishes because of the distortionary and unstable debt environment, tax revenues will fall, as will the ability of the government to service its debt. The possibility of this unpleasant cycle unfolding in highly indebted SIDS is yet another reason to raise concerns over the level of indebtedness in such countries.

Conclusion

This chapter has reviewed the literature on the relationship between public debt and development. A growing consensus among theoretical reasoning and empirical evidence was found to suggest that low to moderate levels of debt can be beneficial to development provided that the debt is properly used, but high levels of debt tend to be deleterious to targeted developmental outcomes. The impact of debt on economic growth and development was manifested through numerous channels, including the fiscal balance, financial sector, macroeconomic policies and institutions, and private investors and entrepreneurs. Complex relationships were found, wherein the channels that caused the buildup of debt turn out to be the same ones through which adverse effects were felt when debt thresholds were crossed.

Notes

*The material in this chapter draws heavily from one by the same author and published as "Chapter 1: Debt and Development: A Review of the Literature" in Caribbean Development Bank (2013).

1. Caner, Grenner and Koehler-Geib (2010).
2. Panizza, Sturzenegger, and Zettelmeyer (2010).
3. Imbs and Ranciere (2005).
4. Cecchetti, Mohanty, and Zampolli (2011, 3).
5. Bourne (2010).
6. As quoted in Deshpande (1997, 171).
7. The DOH is often presented as a Debt Laffer curve, "which posits that larger debt stocks tend to be associated with lower probabilities of debt repayment. On the upward sloping or 'good' section of the curve, increases in the face value of debt are associated with increases in expected debt repayment, while increases in debt reduce expected debt repayment on the downward sloping or 'bad' section of the curve." Pattillo, Poirson, and Ricci (2004).
8. Robinson (2014).
9. Robinson (2014).
10. Todaro and Smith (2006).
11. As quoted in Sheikh, Faridi, and Tariq (2010).
12. Ogunmuyiwa (2011).
13. As quoted in Checherita and Rother (2010, 10).
14. Foss (1997).
15. Deshpande (1997).
16. Scott-Joseph, Melville, Kendall, and Harris (2006).
17. Hanson (2007) notes "(a) many of the largest increases observed in domestic government debt were in crisis countries or those where state banks were recapitalized, and (b) the increase in domestic government debt would have been much less in many countries between the mid-1990s and 2004 had there been no financial crises" (13).
18. Checherita and Rother (2010).
19. Abbas and Christensen (2007).
20. Hanson (2007).
21. Clements, Bhattacharya, and Nguyen (2003, 5) note that "external debt service...can also potentially affect growth by crowding out private investment or altering the composition of public spending. Other things being equal, higher debt service can raise the government's interest bill and budget deficit, reducing public savings; this, in turn, may either raise interest rates or crowd out credit available for private investment, dampening economic growth. Higher debt service payments can also have adverse effects on the composition of public spending by squeezing the amount of resources available for infrastructure and human capital, with negative effects on growth."
22. Checherita and Rother (2010).
23. Abbas and Christensen (2007, 4).

24. They note that this result supports "the hypothesis that institutional and retail participation in the government securities market boosts competition in the financial sector, both on the deposit-taking side—as banks have to compete with government for individual and institutional deposits—and, on the investment side—as banks compete with other sectors in public sector securities auctions. This increased competition should put downward pressure on banks' overheads and intermediation margins, partly alleviating the efficiency concerns associated with high bank holdings of domestic debt." (Abbas and Christensen 2007, 21).
25. Governments can default on domestic debt either "explicitly or through holding down interest rates or other changes in the conditions on domestic debt" (Hanson 2007, 24).
26. Abbas and Christensen (2007).
27. Kirkpatrick and Tennant (2002).
28. Deshpande (1997) notes that after the 1980s debt crisis, many heavily indebted countries (through IMF programs) sought to use exchange rate devaluation to improve their balance of payments position. This resulted in large swings in the current account balances of highly indebted countries, which were achieved through import cuts rather than increases in export revenues. This had adverse repercussions on investments, as many countries relied on imports for a large proportion of capital goods.
29. "This problem would be more acute the greater are the debt servicing obligations, and politically, explanations of the government would be more unacceptable the wider the divergence of interests between the ruling classes and the majority of the population" (Deshpande 1997, 173).
30. "The anticipation of future debt relief needed to resolve the debt overhang problem may also reduce governments' incentive to pursue policy reforms that strengthen their capacity to repay, with a similar negative impact on productivity growth" (Pattillo, Poirson, and Ricci 2004, 6).
31. Abbas and Christensen (2007).
32. Deshpande (1997).
33. Pattillo, Poirson, and Ricci (2004).

References

Abbas, S. M. Ali and Jakob E. Christensen. 2007. "The Role of Domestic Debt Markets in Economic Growth: An Empirical Investigation for Low-income Countries and Emerging Markets." *IMF Working Paper WP/07/127.*

Aschauer, David Alan. 1989. "Is Public Expenditure Productive?" *Journal of Monetary Economics* 23: 177–200.

Bourne, Compton. 2010. "A Discourse on Debt and Economic Growth in the Caribbean Community." www.mona.uwi.edu/economics/notices/Presentations/Comptonpercent20Bourne.pdf.

Caner, Mehmet, Thomas Grenner, and Fritzi Koehler-Geib. 2010. "Finding the Tipping Point: When Sovereign Debt Turns Bad." *World Bank Policy Research Paper WPS5391.*

Cecchetti, Stephen G., M. S. Mohanty, and Fabrizio Zampolli. 2011. "The Real Effects of Debt." http://www.bis.org/publ/othp16.pdf.

Checherita, Cristina and Phillip Rother. 2010. "The Impact of High and Growing Government Debt on Economic Growth: An Empirical Investigation for the Euro Area." *European Central Bank Working Paper Series No 1237.*

Clements, Benedict, Rina Bhattacharya, and Toan Quoc Nguyen. 2003. "External Debt, Public Investment, and Growth in Low-Income Countries." *IMF Working Paper WP/03/249.*

Cordella, Tito, Luca Antonio Ricci, and Marta Ruiz-Arranz. 2005. "Debt Overhang or Debt Irrelevance? Revisiting the Debt-Growth Link." *IMF Working Paper WP/05/223.*

Deshpande, Ashwini. 1997. "The Debt Overhang and the Disincentive to Invest." *Journal of Development Economics* 52: 169–187.

Foss, Nicolai J. 1997. "The New Growth Theory: Some Intellectual Growth Accounting'." Department of Industrial Economics and Strategy, *Working Paper 97–2*, http://ep.lib.cbs.dk/download/ISBN/8778690072.pdf.

Fosu, Augustin Kwasi. 2007. "The External Debt-Servicing Constraint and Public Expenditure Composition." *UNU-WIDER Research Paper No. 2007/36.*

Hanson, James A. 2007. "The Growth in Government Domestic Debt: Changing Burdens and Risks." *World Bank Policy Research Working Paper WPS4348.*

Husain, Aasim M. 1997. "Domestic Taxes and the External Debt Laffer Curve." *Economica* 64(255): 519–525.

Imbs, Jean and Romain Ranciere. 2005. "The Overhang Hangover." Paper presented at CEPR Conference on Institutions, Policy and Growth, INSEAD, May. Washington, DC: World Bank.

Kemal, Abdul Razzaq. 2001. "Debt Accumulation and Its Implications for Growth and Poverty." *The Pakistan Development Review* 40(4): 263–281.

Khan, Moshin and Kumar, Manmohan. 1997. "Public and Private Investment and the Growth Process in Developing Countries." *Oxford Bulletin of Economics and Statistics* 59(1): 69–88.

Kirkpatrick, Colin and Tennant, David. 2002. "Responding to Financial Crisis: The Case of Jamaica." *World Development* 30(11): 1933–1950.

Krugman, Paul. 1988. "Financing vs Forgiving: A Debt Overhang." *Working Paper 2486, National Bureau of Economic Research.*

Ogunmuyiwa, Michael S. 2011. "Does Fiscal Deficit Determine the Size of External Debt in Nigeria?" *Journal of Economics and International Finance* 3(10): 580–585.

Panizza, Ugo, Federico Sturzenegger, and Jeromin Zettelmeyer. 2010. "International Government Debt." *UNCTAD Discussion Papers 199 (June).*

Pattillo, Catherine, Helene Poirson, and Luca Ricci. 2004. "What are the Channels through which External Debt Affects Growth?" *IMF Working Paper WP/04/15.*

Reinhart, Carmen M. and Kenneth S. Rogof. 2010. "Growth in a Time of Debt." *NBER Working Paper Series 15639.*

Robinson, Michele. 2014. "The Debt Experience of SIDS in the Caribbean." In *Debt and Development in Small Island Developing States*, edited by Damien King and David Tennant. Basingstoke: Palgrave McMillan.

Sahay, Ratna. 2006. "Stabilization, Debt and Fiscal Policy in the Caribbean." In Ratna Sahay, David O. Robinson, and Paul Cashin (eds), *The Caribbean: From Vulnerability to Growth (International Monetary Fund).*

Schclarek, Alfredo and Franciso Ramon-Ballester. 2005. "External Debt and Economic Growth in Latin America." http://www.cbaeconomia.com/Debt-latin.pdf.

Scott-Joseph, Ankie, Juliette Melville, Patrick Kendall, and Elmer Harris. 2006. "Debt Sustainability in Caribbean Countries: An Exploration of Alternative Methodologies." St. Michael: Social and Economic Research Unit, Caribbean Development Bank.

Sheikh, Muhammad Ramzan, Muhammad Zahir Faridi, and Khadija Tariq. 2010. "Domestic Debt and Economic Growth in Pakistan: An Empirical Analysis." *Pakistan Journal of Social Sciences* 30(2): 373–387.

Todaro, Michael P. and Stephen C. Smith. 2006. *Economic Development – 9th Edition*. Essex, UK: Pearson Education.

Uzochukwu, Amakom. 2003. *Nigeria Public Debt and Economic Growth: An Empirical Assessment of Effects on Poverty*. Enugu Nigeria: African Institute for Applied Economics.

Chapter 3

The Debt Experience of SIDS in the Caribbean

Michele Robinson

Overview of the Caribbean

The Caribbean[1] comprises 13 independent nation states: Antigua and Barbuda, The Bahamas, Barbados, Belize, Dominica, Grenada, Guyana, Jamaica, St. Kitts and Nevis, St. Lucia, St. Vincent and the Grenadines, Suriname, and Trinidad and Tobago. With populations of less than 1.5 million,[2] all are defined as small states. There are ten Caribbean island states as distinct from the mainland states of Belize, Guyana, and Suriname. These ten countries, for purpose of analysis, comprise the Caribbean small island developing states (SIDS). Table 3.1 presents a number of indicators that highlight some of the key characteristics of these countries.

Caribbean SIDs can be divided into two regional groupings, namely, the six member countries and "micro states" [3] of the Eastern Caribbean Currency Union (ECCU) and the four non-ECCU countries comprising The Bahamas, Barbados, Jamaica, and Trinidad and Tobago. The ECCU member states share a common currency, the Eastern Caribbean (EC) dollar, which is pegged to the US dollar, as well as a single central bank, the Eastern Caribbean Central Bank (ECCB).

As indicated in table 3.2, Caribbean SIDS are predominantly upper middle-income countries with an average per capita income in 2012 of US$11,812 for all countries. However, this average masks widely disparate per capita income levels ranging from a low of US$5,541 in Jamaica to a high of US$22,833 in The Bahamas. Since 2010 and based on their growth in per capita incomes, three countries—Antigua and Barbuda, Barbados, and St. Kitts and Nevis—have graduated from their income class to be reclassified as high-income countries along with The Bahamas and Trinidad and Tobago. The change in their income status

Table 3.1 Caribbean small island states: Selected social and economic indicators

Country	Geography	Environmental Vulnerability 1/	Population in millions (2012)	Distance from the United States (in miles)	2012 GNI per capita (current US$)	Income Classification	2011 Human Development Index Ranking (out of 187 countries)	Public debt-to-GDP (in percent) (2012)
Island States								
Antigua and Barbuda	island	Vulnerable	0.089	1,337	13,429	High	67	97.8
The Bahamas	island	At risk	0.372	112	22,832	High	49	52.6
Barbados	island	Extremely vulnerable	0.283	1,611	16,151	High	38	72.6
Dominica	island	Extremely vulnerable	0.071	1,414	7,022	Upper middle	72	72.3
Grenada	island	Extremely vulnerable	0.106	1,567	7,496	Upper middle	63	112.6
Jamaica	island	Extremely vulnerable	2.712	524	5,541	Upper middle	85	143.3
St. Kitts and Nevis	island	Highly vulnerable	0.054	1,275	12,804	High	72	144.9
St. Lucia	island	Extremely vulnerable	0.181	1,496	7,276	Upper middle	88	78.7
St. Vincent and the Grenadines	island	Highly vulnerable	0.109	1,337	6,489	Upper middle	83	68.3
Trinidad and Tobago	island	Extremely vulnerable	1.337	1,622	9,018	High	67	35.7
Mainland States								
Belize	mainland	At Risk	0.324	767	4,535	Upper middle	96	81.0
Guyana	mainland	Resilient	0.795	1,622	3,596	Lower middle	118	60.4
Suriname	mainland	Resilient	0.535	1,360	7,997	Upper middle	105	18.6
Caribbean				**1,234**			77	**79.9**

Source(s): International Monetary Fund (2013d), United Nations Development Programme (2011) and World Bank (2012).

Table 3.2 Caribbean small states: 2012 per capita income

Income Classification		Country	Per Capita Income 2012 (US$ current prices)
High-Income Countries (US$12, 616 and above)	Island States	The Bahamas	22,833
		Trinidad and Tobago	19,018
		Barbados	16,152
		Antigua and Barbuda	13,428
		St. Kitts and Nevis	12,804
Upper-Middle-Income Countries (US$3, 946 – $12, 195)	Island states	Grenada	7,497
		St. Lucia	7,276
		Dominica	7,022
		St. Vincent & the Grenadines	6,489
		Jamaica	5,541
	Mainland	Suriname	8,686
		Belize	4,536
Lower-Middle-Income Countries (US$1,036 – $4,085)	Mainland	Guyana	3,596

Source(s): International Monetary Fund (2013c), World Bank (2013).

constrains these countries' ability to access concessional aid flows from abroad.

Socioeconomic development varies considerably across Caribbean SIDS as indicated by the United Nations (UN) Human Development Index (HDI). Barbados places highest in human development with an index ranking of 38 out of a total of 187 countries. St Lucia ranks lowest with a ranking of 88, just behind Jamaica which ranks 85th. Caribbean SIDS nonetheless fare better than their mainland counterparts where Guyana and Suriname are the only two Caribbean small states to rank below 100.

Despite these differences, Caribbean SIDS share distinct commonalities. Given their small size and geographical location, most Caribbean SIDS are extremely vulnerable to natural disasters, especially earthquakes and hurricanes (see table 3.3). In addition to their vulnerability to natural disasters, the economies of Caribbean SIDS are challenged by their small size, narrow economic base, and openness to trade. Reflective of their small size, Caribbean SIDS have limited ability to enjoy economies of scale in production (International Monetary Fund 2013a). As a result, production costs tend to be higher and markets concentrated in only a few sectors.

Caribbean SIDS, with the exception of Trinidad and Tobago, are all tourism-dependent economies. Trinidad and Tobago is the sole

Table 3.3 Caribbean SIDS: Classification by environmental vulnerability

Extremely vulnerable	Highly Vulnerable	Vulnerable	At Risk	Resilient
Barbados Grenada Jamaica St. Lucia Trinidad and Tobago	Dominica* St. Kitts and Nevis St. Vincent & the Grenadines	Antigua and Barbuda	The Bahamas	

Notes: *There were no data available for Dominica in the EVI. However, based on its size and location in relation to its comparators, it is considered to be highly environmentally vulnerable.

Source(s): UNEP/South Pacific (SOPAC) Allied GeoScience Commission (2012).

commodity exporter, relying primarily on its exports of oil and natural gas. Six of the nine tourism-dependent countries—Antigua and Barbuda, The Bahamas, Barbados, Grenada, St. Kitts and Nevis, and St Lucia—are further classified as tourism-intensive economies, as more than 50 percent of their GDP is generated from tourism. A few tourism-dependent countries are providers of other services, mainly in the offshore banking sector. However, despite the growth in this sector, tourism remains the primary economic mainstay of all these countries.

A key feature of Caribbean SIDS is their openness to international trade. With a narrow production base, and exports limited to a few goods and services; import dependence is high, particularly for foods, fuel, and other raw materials. In 2011, merchandise trade as a percentage of GDP amounted to an average 55 percent for tourism-dependent Caribbean SIDS. This indicator stood at 68 percent for St Lucia. Trinidad and Tobago with merchandise trade-to-GDP of 33 percent in 2011 lowers the Caribbean average to 52.8 percent. Trinidad and Tobago's relatively low measure of trade openness reflects the oil state's lower reliance on imports (particularly fuels) and greater diversification in its domestic economy. Overall, however, trade openness is a significant factor in Caribbean SIDS' vulnerability to external shocks.

Overview of Public Debt of SIDS in the Caribbean

Caribbean SIDS are among the most heavily indebted countries in the world. At the end of 2012, half of the ten most highly indebted middle-income countries were Caribbean SIDS (figure 3.1). Three of these countries were among the top five, with St. Kitts and Nevis topping the list with total public debt-to-GDP of 144.9 percent followed closely by Jamaica with debt-to-GDP of 143.3 percent.

Of the ten Caribbean SIDS, all 6 ECCU member states are included among the top 15 most highly indebted countries in the world. Half

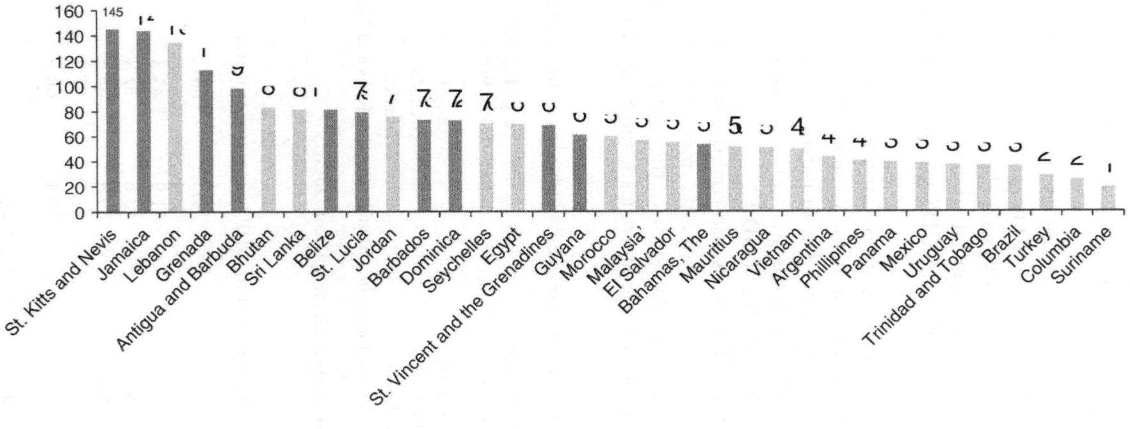

Figure 3.1 Public debt-to-GDP of middle-income countries, as at end 2012 (in %)

of the ECCU states are classified as very highly indebted and the other half as highly indebted, all registering debt-to-GDP levels in excess of 60 percent—the 2020 sustainability target agreed upon by all ECCU member states (see table 3.4).

Only one non-ECCU country—Jamaica—is classified as very highly indebted with debt-to-GDP exceeding 140 percent at the end of 2012. Barbados, which is highly indebted, ranks second among the non-ECCU countries with debt-to-GDP of 72.6 percent. Trinidad and Tobago and the Bahamas are the only two Caribbean SIDS with public debt-to-GDP below 60 percent.

The average public debt-to-GDP levels of Caribbean SIDS far exceed the average for all middle-income countries. Compared to a 2012 average debt-to-GDP ratio of 55.6 percent for all indebted middle-income countries, Caribbean SIDS' debt levels were over 50 percent higher, amounting to an average 87.9 percent of GDP.

Caribbean SIDS' debt-to-GDP ratios also compare unfavorably with other small states (Table 3.5). At the end of 2012, average public debt-to-GDP among Caribbean SIDS was almost twice the average of the small states of the Africa, Indian Ocean, Mediterranean, and South China Sea

Table 3.4 Caribbean SIDS: Classification of public indebtedness–2012 (end period)

Low to Medium Debt (0% to 60%)	High Debt (Over 60% to 90%)	Very High Debt (Over 90%)
The Bahamas	Barbados	Antigua and Barbuda
Trinidad and Tobago	Dominica	Grenada
	St. Lucia	Jamaica
	St. Vincent and the Grenadines	St. Kitts and Nevis

Source(s): International Monetary Fund (2013c) World Bank (2013).

Table 3.5 SIDS: Average debt-to-GDP by regional grouping

Region	2011	2012
AIMS* small states	41.9	43.3
Pacific** small states	43.6	37.8
Caribbean small states	84.9	85.3
Caribbean SIDS	85.7	87.9
All Small States	60.4	60.5

Source(s): International Monetary Fund (2013c).

Notes: *AIMS refer to the Africa, Indian Ocean, Mediterranean and South China Sea region. The small states included in this sample include Botswana, The Gambia, Lesotho, Mauritius, Namibia, Seychelles, and Swaziland in Africa and the Maldives in Asia.

** The Pacific states in this table include Fiji, Kiribati, Papua New Guinea, Samoa, Tonga, Tuvalu and Vanuatu.

(AIMS) small states (43.3 percent) and more than twice as high as the average debt-to-GDP ratio of Pacific small states (37.8 percent).

The significant divergence in levels of indebtedness especially when compared to African small developing states is to some extent explained by significant debt reductions achieved under the Highly Indebted Poor Countries (HIPC) initiative. It also reflects a greater reliance by the African states on concessional financing and grants for budgetary and balance of payments support. In addition, most African small states rely on commodity exports from either the mineral or agricultural sector rather than the services sector such as tourism or banking as in the case of the Caribbean. Higher export earnings and tax revenues, and stronger growth have helped to keep debt, on average, at lower levels. Another significant factor is that AIMS small states with a single exception, the Seychelles, are far less vulnerable to natural disasters than their Caribbean peers and data indicate their fiscal consolidation efforts are generally stronger in the aftermath of a disaster.

Caribbean SIDS have also underperformed when compared to Pacific small states, which have maintained relatively moderate ratios of public debt-to-GDP since the 2000s. At the end of 2012, the median value of public debt-to-GDP for Pacific small states was 44.3 percent, with the highest debt-to-GDP recorded in Samoa of 80.3 percent. Similar to Caribbean SIDS, all Pacific states are island states, have similarly open economies and narrow production bases and are highly vulnerable to natural disasters. Pacific states face the added disadvantage of remoteness with long distances between themselves and their major trading partners. However, while Caribbean SIDS have relied heavily on foreign direct investment (FDI) and debt financing, Pacific states have been financed mainly through official grants (Browne 2006). Thus while Pacific states have managed to keep their external and fiscal positions at manageable levels and faced moderate levels of indebtedness, Caribbean SIDS have recorded much weaker fiscal and external balances and higher levels of indebtedness. While 7 of 8 Pacific small states had public debt-to-GDP below 60 percent at the end of 2011, only 2 of 10 Caribbean countries, The Bahamas and Trinidad and Tobago, were below this widely accepted sustainability threshold.

A significant distinction between Caribbean SIDS and AIMS/Pacific small states is their access to the international capital markets. While only a few AIMS/Pacific small states have a sovereign credit rating, over half of the ten Caribbean SIDS have assigned credit ratings from one or more of the major international credit rating agencies, Moody's Investor Services, Standard and Poor's, or Fitch (see table 3.6).

As at the end of 2012 only two Caribbean SIDS, The Bahamas and Trinidad and Tobago enjoyed investment grade status on their long-term foreign currency bonds. However, after years of an A rating, in 2012,

Table 3.6 Caribbean SIDS: Long-term foreign-currency ratings, 2006, 2012 (end period)

	Moody's		Standard and Poor's	
	2006	2012	2006	2012
1. Trinidad and Tobago	Baa1	Baa1	A–	A
2. The Bahamas	A3	Baa1	A–	BBB
3. Barbados	Baa2	Ba1	BBB+	BB+
4. St Vincent and the Grenadines	B1*	B2	..	..
5. Jamaica	B3	B3	B–	CCC+
6. Belize	Caa3	Ca	CC	B–
7. Grenada	..	..	B–	SD

Notes: *This rating for St. Vincent and the Grenadines is at December 10, 2007, when the country was first rated.

Source(s): Moody's Investor Services, Standard and Poor's Rating Services.

Table 3.7 Caribbean SIDS: Number of ratings downgrades, 2001–2005, 2006–2010, 2011–2012

	Moody's			Standard and Poor's		
	2001–5	2006–10	2011–12	2001–5	2006–10	2011–12
1. Trinidad and Tobago	0	0	0	0	0	0
2. The Bahamas	0	0	1	0	1	1
3. Barbados	0	1	1	1	2	1
4. St Vincent and the Grenadines	n.r.	0	1	n.r.	n.r.	n.r.
5. Jamaica	1	2	0	1	3	0
6. Belize	4	0	2	5	1	4
7. Grenada	n.r.	n.r.	n.r.	2	1	1

Notes: n.r. – not rated.

Source(s): Moody's Investor Services (2012) and Standard and Poor's Rating Services (2012).

the Bahamas was downgraded to the lower end of the investment class, falling from A3 to Baa1.

As shown in table 3.7, since the 2000s, an increasing number of Caribbean SIDS have been downgraded, the frequency increasing in the latter part of the decade following restructuring operations in Belize, Grenada, and Jamaica, and weakening fiscal and external positions and high and rising debt in Barbados and St. Vincent and the Grenadines. Amid dwindling external concessional aid flows, the speculative grade rank and slide in ratings for most Caribbean SIDS has translated into relatively high costs of borrowing, adding to their already high debt burdens.

The Debt Profile of Caribbean SIDS

Evolution of the Debt

Almost all Caribbean SIDS experienced a buildup in their debt over the period 2000–2003 with the rate of accumulation most pronounced in the year immediately after the September 11, 2001 terrorist attacks in the United States. As highlighted in figure 3.2, Caribbean debt jumped by 12 percentage points from an average of 71.8 percent at the end of 2001 to 83.7 percent at the end of 2002. Caribbean SIDS public debt climbed even further to peak at 88.9 percent of GDP at the end of 2003. The onset of the global financial–food–fuel crisis in 2008 also brought a significant upturn in debt-to-GDP ratios among Caribbean SIDS. A year into the global financial crisis, at end 2009, Caribbean SIDS recorded a year-on-year increase in public debt-to-GDP of 16.4 percent, one of the largest average annual increases for the decade and second only to the 16.5 percent annual increase recorded at the end of 2002, in the aftermath of the September 11, 2001 attacks on the United States.

Declines in average debt-to-GDP among some Caribbean SIDS, namely Antigua and Barbuda, and Dominica were largely reflective of severe payments difficulties that prompted comprehensive debt restructuring operations in both countries as well as a transitory recovery in economic activity in the aftermath of the September 11, 2001 terrorist attacks.

Figure 3.3 shows that the highest levels of public debt-to-GDP have been recorded in St. Kitts and Nevis and Jamaica, both with ratios exceeding 100 percent since 2001 and 140 percent since 2009. Similar to Antigua and Barbuda, both countries have had a long history of debt accumulation which preceded the 2000s.

The Bahamas is only one of two Caribbean SIDS that has remained moderately indebted over 2000–2012. However, public debt has been on a steady upward trajectory in the second half of the 2000s since the 2008 global financial crisis devastated the country's tourism sector, leading to economic contraction. The concomitant drop in tourism receipts impacted tax revenues and strained the public finances. In just one year the fiscal deficit doubled from 1.8 percent of GDP to 3.9 percent of GDP. The public debt rose from 33 percent of GDP in 2008 to 38 percent of GDP by 2009 and stood at 52.6 percent at end 2012. As highlighted in figure 3.4, the growth in debt from 2001 to 2012 has been the largest among Caribbean SIDS in percentage terms increasing by over 115 percent.

Debt Service

External debt service ratios in excess of 15 percent of exports are said to be unsustainable. Similarly, total debt service ratios, including external

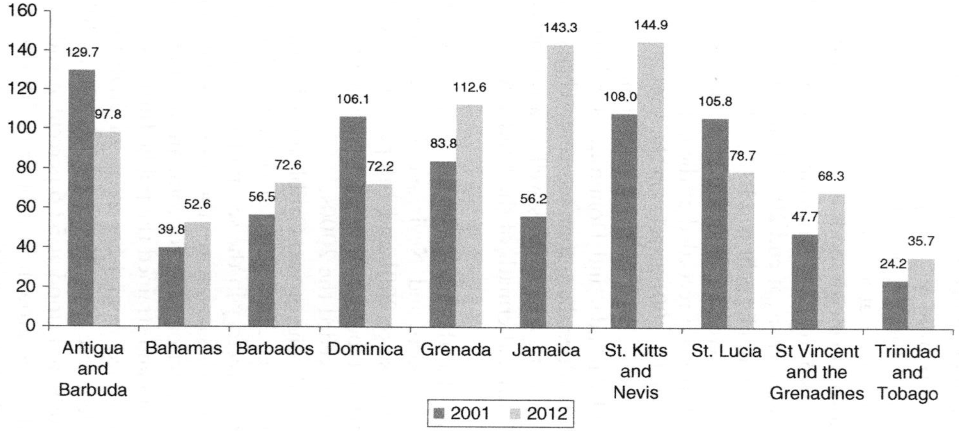

Figure 3.2 Caribbean SIDS: Public debt-to-GDP, 2001 and 2012 (in %, end period)

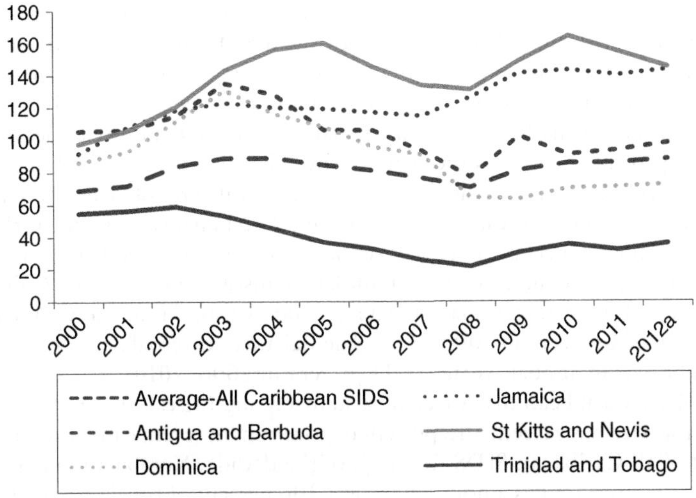

Figure 3.3 Caribbean SIDS: Public debt-to-GDP by sub-periods

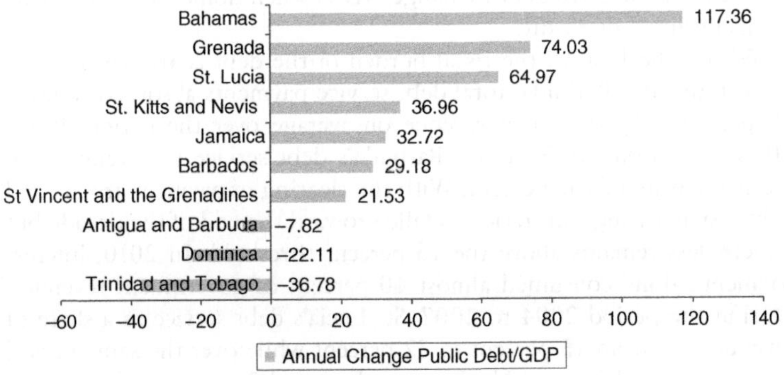

Figure 3.4 Caribbean SIDS: Public debt-to-GDP, end 2012 over end 2001 (percentage change)

Source(s): International Monetary Fund (2013c).

and domestic debt, greater than 15 percent of government revenues, breach sustainability thresholds. Largely reflective of concessional terms received from official donors, external debt service ratios in Caribbean SIDS have kept relatively low. On average, external debt service ratios amounted to around 10 percent in the period 2001–2005, just creeping up marginally to 11 percent over the second half of the decade. Grenada had the highest external debt service ratios averaging 21 percent for

2001–2005 and 24 percent in 2006–2012 on the basis of a relatively high share of external commercial borrowings. Even with considerable borrowing from the international capital markets in the early half of the decade, Jamaica's external debt service ratio averaged less than 12 percent in the first half of the decade. Jamaica's external debt service peaked at 17 percent in 2005 as the share of sovereign bonds increased but has since declined to an average 11 percent as multilateral obligations have increasingly replaced capital market issues. The Bahamas has shown the most dramatic increase in its debt service ratio over the period as external borrowings have increased from multilateral institutions and from China in late 2000s (International Monetary Fund 2013b). From a low annual average of 4 percent in 2001–2005, The Bahamas' external debt ratio has climbed to an annual average of 11 percent in 2006–2010.

Many Caribbean SIDS face unsustainably high fiscal burdens. Total debt service has exceeded 15 percent of government revenue in more than half of all Caribbean SIDS for much of the decade. Notably in Jamaica, total debt service has amounted to over 100 percent of fiscal revenue for most of the 2000s. From a peak of 157 percent in 2001, debt service-to-revenue has declined to 67 percent in 2012. The fall in the ratio occurred mainly after the 2010 debt exchange (JDX) when domestic interest payments declined markedly.

Within the ECCU, the fiscal burden of the debt is also substantial. In Antigua and Barbuda, total debt service payments absorbed roughly 50 percent of government revenue on average over the period 2001–2005.[4] In 2005, Antigua and Barbuda's debt service to revenue ratio reached a peak of 107 percent. With the clearing of payment arrears and debt restructuring, this ratio has fallen toward the end of the decade but nonetheless remains above the 15 percent threshold. In 2010, interest payments alone consumed almost 40 percent of government revenues. During the period 2004 to 2007, St. Lucia's debt service as a share of revenue rose from 15 percent to 32 percent while over the same period the ratio in St. Kitts and Nevis hovered around 35 percent. St. Vincent and the Grenadines' fiscal burden, albeit lower than its comparators, has averaged around 20 percent for the latter half of the 2000s.

Composition of the Debt

Domestic versus External

The majority of Caribbean SIDS are domestically indebted. Of the ten Caribbean SIDS, all the non-ECCU countries, as well as Antigua and Barbuda and St. Kitts and Nevis, have high shares of domestic debt in their public debt portfolio (see table 3.8). However, the majority of ECCU countries—Dominica, Grenada, St. Lucia, and St. Vincent and the Grenadines, have traditionally relied heavily on overseas financing as reflected in their high shares of external debt.

Table 3.8 Caribbean SIDS: Composition of public debt, domestic and external, as at end 2012

	Domestic Debt > 50% of total	External Debt > 50% of total
Highly indebted (Debt/GDP > 60%)	Antigua and Barbuda Barbados Jamaica St. Kitts and Nevis	Dominica Grenada St. Lucia St. Vincent and the Grenadines
Moderately indebted (Debt/GDP 30–60%)	Bahamas, The Trinidad and Tobago	
Less Indebted (Debt/GDP < 30%)		

Source(s): International Monetary Fund (2004, 2007, 2009).

Among ECCU countries, Dominica has had the highest share of external debt with close to 80 percent of total public debt owed to external creditors. Dominica's external share has been growing over the period 2000–2012. From a share of just over 60 percent at the end of 2000, Dominica's external debt has increased to 80 percent of the total, with the bulk of the obligations owed to multilateral development banks, particularly the Caribbean Development Bank (CDB). The growth in Dominica's debt reflects the buildup in domestic arrears in the early half of the decade and, amid steadily deteriorating finances, the increased reliance on multilateral financial support. Because Dominica's external borrowings were mainly US dollar obligations, much of the attendant foreign currency risk exposure was mitigated given the fixed-peg arrangement between the Eastern Caribbean and the US dollar.

In other high external debt ECCU countries—especially St. Lucia and St. Vincent and the Grenadines—the share of external public debt has perceptibly fallen over the period 2000–2012. Dwindling external aid flows to the region have been a principal factor behind the decline. Even where external donors have pledged assistance, disbursements have fallen far short of commitments. For example, over the period 2000–2006, little of the pledged European Union support, for Caribbean countries affected by the dismantling of trade preferences for sugar and bananas, materialized. Of the €40.8 million committed to Dominica over the period, only €5.1 million or 12.5 percent was disbursed at the end of 2006. Similarly, of the €52.1 million and €38.3 million pledged to St. Lucia, and St. Vincent and the Grenadines respectively, only €10.6 million (20.3 percent) and €1.8 million (17.0 percent) were disbursed. In the absence of substantial foreign financing, both ECCU countries and other Caribbean SIDS have increased their borrowings in the domestic markets, leading to a concomitant rise in the domestic debt shares in their debt portfolios. Shares of external debt among the more externally indebted ECCU countries, excluding Dominica and Grenada, fell from an average 76 percent of total public debt at the end of 2000 to 64 percent at end 2011.

With public debt at 143 percent of GDP at end 2012, Jamaica is the most heavily indebted of the non-ECCU Caribbean SIDS. More than half of the public debt is owed to domestic creditors. Jamaica's debt composition began shifting in the 1990s when, after exiting Paris Club reschedulings, the authorities began funding their budget domestically rather than from external sources. From a share of just 16 percent at that 1993 turning point, domestic debt as a percentage of total debt more than tripled, reaching 55 percent at the end of 2000. This share increased to 62 percent in 2003 when financial sector liabilities were assumed on the central government balance sheets but subsequently tapered to an average share of 56 percent for the remainder of the decade.

St. Kitts and Nevis, like Jamaica, is also mainly obligated to domestic lenders. However, whereas Jamaica holds its domestic debt primarily in the form of marketable securities, holders of St. Kitts domestic debt mainly comprise indigenous commercial banks as well as the public corporations such as the Social Security Board (International Monetary Fund 2012). The substantial claims by the banking sector raise serious issues of financial sector stability in the event of a debt restructuring operation or in the event of a severe payments crisis. At the end of 2011, domestic debt comprised more than 70 percent of total public debt and over 95 percent of GDP.

External Debt—Official versus Private Creditors

Since the start of the decade, access to the international capital markets has been enjoyed by all non-ECCU countries as well as two of the six ECCU member states (see table 3.9). The Bahamas, Barbados, Jamaica, and Trinidad and Tobago have all issued sovereign bonds in the international capital markets. Of the four countries, Jamaica has accessed the capital markets most frequently and for the first half of the decade, sovereign bond issues accounted for over 60 percent of the external portfolio. With the tightening of credit conditions in the capital markets and a series of ratings downgrades, Jamaica, in the second half of the decade, took a deliberate shift in policy toward relying more heavily on less costly multilateral aid in lieu of financing from the capital markets. At the end of 2012, multilateral debt accounted for a 40 percent share of total external debt compared to 27 percent in 2001, while bonded debt accounted for 51 percent of total external debt at the end of 2012.

Table 3.9 Caribbean SIDS: Public indebtedness by market access, as at end 2012

	Market Access	Non-Market Access
Highly indebted (Debt/GDP > 60%)	Barbados Grenada Jamaica St. Vincent and the Grenadines	Antigua and Barbuda Dominica St. Lucia St. Kitts and Nevis
Moderately indebted (Debt/GDP 30 – 60%)	Bahamas, The Trinidad and Tobago	
Less Indebted (Debt/GDP < 30%)		

Source(s): International Monetary Fund (2013c).

ECCU member states owe external debt primarily to multilateral creditors (an average of 40 percent). With limited access to the international capital markets and dwindling bilateral aid flow, borrowings from external sources have either been from commercial banks or from multilateral financial institutions. Funds have been primarily used to finance projects under government public sector investment programs. Increasingly, the World Bank and Caribbean Development Bank, as well as the International Monetary Fund, have been relied on for emergency financing in the wake of a natural disaster.

Risk Exposures in SIDS Portfolio Composition

Many Caribbean SIDS are highly vulnerable to refinancing risk associated with the high share of short-term domestic debt (Treasury bills and overdraft facilities). Foreign currency risk among ECCU countries is low as their exchange rate is fixed to the US dollar and US-dollar credits comprise the bulk of external borrowings. This currency share is changing, however, as China lends more to the region and reliance on IMF assistance increases the currency share denominated in Special Drawing Rights.

Among non ECCU countries, Jamaica's debt portfolio is perhaps the most risky. The portfolio comprises a high share of foreign currency debt contracted both overseas as well as in the domestic market. Some 20 percent of Jamaica's domestic debt is denominated in foreign currency and overall, foreign currency debt comprised more than 55 percent of the total debt portfolio. Floating rate debt is also significant accounting for 34 percent of domestic debt at the end of 2012, down from 60 percent at the end of 2008. While external debt is primarily long-term, up to 2011, over 55 percent of domestic debt instruments had a maturity of less than five years. This share fell to 23 percent after the 2013 domestic debt exchange.

Managing the portfolio composition is essential in keeping costs and risks within prudent levels. Even with limited borrowing options, Caribbean SIDS have increasingly become aware of the need to borrow more strategically. Many have developed medium-term debt strategies to identify available markets and to assess the impact on borrowing costs from accessing funds from various lending sources or in particular currencies, instruments, and maturities.

Current Factors Affecting Debt Trends in the Caribbean SIDS

Caribbean SIDS have had persistently high levels of debt throughout the 2000s. Public debt to GDP levels have averaged over 100 percent over this period and debt service as a proportion of fiscal revenues has far exceeded the 15 percent sustainability threshold, surpassing 100 percent in a number of countries. The increases in public debt and the growing unsustainability have been occasioned by both domestic policies, which generally have been fiscally accommodative, and a series of external shocks both economic and weather-related.

Dismantling of EU Trade Arrangements

The steady erosion of European preferential trade arrangements for sugar and bananas was a significant factor in the debt accumulation witnessed by Caribbean SIDS particularly in ECCU states. The dismantling of preferential access began in 1993, when the European Union removed internal trade barriers exposing Caribbean sugar and banana exporters to increased competition from nontraditional suppliers. In 1998, the EU eliminated banana quotas enabling access to the once protected EU and UK market by more efficient low-cost African and Latin American producers. Between 2000 and 2002, the share of ECCU[5] banana imports into the United Kingdom fell by almost half, from 21 percent of total market share to 12 percent. In addition to reduced preferential access for Caribbean bananas, the EU also implemented a phased reduction of the guaranteed price for sugar beginning in 2005. These trade reforms had a severe negative effect on the trade balance, fiscal balance, and economic output of Caribbean SIDS, particularly in ECCU countries, whose economic mainstay for decades had been sugar and banana exports.

ECCU states suffered significantly weakened external positions. Over the period 2001 and 2005, ECCU external current account deficits averaged an annual 15.5 percent of GDP, significantly higher than the Caribbean SIDS regional average of –10.6 percent, or the average among non-ECCU states of 2.2 percent. Dominica, St. Kitts and Nevis,

Table 3.10 Selected Caribbean SIDS: External account balance, 2000–2006 (in percent of GDP)

Country	2000	2001	2002	2003	2004	2005	2006
Dominica	(18.4)	(15.8)	(14.5)	(15.6)	(16.0)	(21.0)	(12.9)
St. Kitts and Nevis	(16.6)	(24.1)	(28.7)	(27.3)	(16.2)	(14.9)	(14.1)
St. Lucia	(12.4)	(15.2)	(14.7)	(18.9)	(10.6)	(14.3)	(30.4)
St Vincent and the Grenadines	(5.9)	(8.3)	(9.1)	(16.5)	(19.6)	(18.0)	(19.3)
All ECCU states	**(11.5)**	**(13.6)**	**(15.1)**	**(17.4)**	**(13.1)**	**(18.2)**	**(21.8)**
All non-ECCU states	**(2.8)**	**(1.9)**	**(5.4**	**(1.8)**	**(1.0)**	**(0.7)**	**1.7**

Source(s): International Monetary Fund (2013c).

and St. Lucia were the worst affected as exports of bananas dropped sharply. Both Dominica and St. Kitts and Nevis realized annual fiscal losses equivalent to 3 to 4 percent of GDP (Sahay 2005) resulting in a rapid widening of fiscal deficits.

Governments in the affected countries pursued accommodative fiscal policies to transition out of banana and sugar production and maintain employment. In closing its sugar industry in 2005, St Kitts and Nevis provided a severance package to its over 1,400 sugar workers amounting to some 2.3 percent of GDP (International Monetary Fund 2007b). In addition, it assumed the servicing of debt for the Sugar Manufacturing Company, a public sector entity, increasing the central government's interest bill by 1.5 percent of GDP per annum. Unsurprisingly, St. Kitts and Nevis public debt became increasingly unsustainable at levels of 160 percent of GDP at end 2005.

Despite the fiscal losses in Dominica, the authorities sought to reinvigorate growth by entertaining large capital expenditures. The overall public sector deficit amounted to 9 percent of GDP, financed primarily through heavy external borrowing and arrears accumulation. The debt burden was exacerbated by the government assuming the debt obligations of the Dominica Banana Marketing Corporation. By 2003, Dominica's public debt-to-GDP reached a historic high of 131 percent of GDP while external debt service as a share of exports of goods and services stood at 9 percent. In 2002, the Dominican authorities requested assistance from the International Monetary Fund under a one-year Stand-By Arrangement and by late 2003 announced its intention to undertake a preemptive restructuring ahead of a likely default.

9/11 Terrorist Attacks

An immediate effect of the September 11, 2001, terrorist attacks in the United States was to severely disrupt tourist arrivals to the Caribbean.

Table 3.11 Performance of Caribbean tourism-dependent SIDS pre- and post-9/11 terrorist attacks

	Year immediately prior to terrorist attack (2000)	Year of terrorist attack (2001)	Year immediately after terrorist attack (2002)
Tourist Arrivals (in thousands)	472.5	457.1	521.4
Real GDP (Annual average %)	3.2	(0.2)	2.2
External Current Account/GDP (%)	(6.9)	(10.4)	(12.6)
Overall fiscal balance/ GDP (%)	(5.2)	(5.6)	(8.4)
Public Debt/GDP (%)	70.5	73.5	86.4

Source(s): International Monetary Fund (2013c), Eastern Caribbean Central Bank (2012), and Caribbean Tourism Organisation (CTO) (2011).

Among tourism-dependent Caribbean SIDS, the impact was swift and severe (see table 3.11). Tourism arrivals fell by 4 percent among tourism-dependent economies declining even further by 6 percent in ECCU countries. Net tourism earnings in ECCU countries fell by 7.5 percent.

Economic activity slumped in all Caribbean SIDS. Compared to an average annual growth rate of 3.6 percent in 2000, economic growth came to a virtual standstill amounting to 0.3 percent in 2001. Already high public debt-to-GDP ratios jumped, as the decline in tourism earnings widened current account deficits and depressed tax revenues prompting governments to borrow to staunch the budgetary and current account deficits.

The 2001 fall in real GDP was most severe in Antigua and Barbuda and St. Lucia where tourism and tourism-related activities account for more than 60 percent of economic activity. Both countries' economies contracted, registering declines of 4.4 percent and 4.8 respectively, following growth rates of 3.3 percent and 1.0 percent in 2000. Public debt as a percentage of GDP in Antigua and Barbuda ratcheted up to 84 percent in 2002 from 69 percent in 2000, reflecting the combination of negative economic growth and increased borrowing. A buildup of debt arrears also occurred as financing debt servicing obligations out of revenues proved increasingly difficult. In St. Lucia, public debt-to-GDP rose by 50 percent, as the ratio jumped from 40 percent to 60 percent. The fiscal deficit almost tripled, increasing to 3.9 percent in 2001 from 1.4 percent a year earlier, reflecting the significant decline in tourism-related tax revenues. Dominica experienced the steepest buildup in debt, with debt-to-GDP of 111 percent in 2002 compared to 86 percent in

2000. The unsustainable debt overhang precipitated a comprehensive debt restructuring exercise in 2004.

Among non-ECCU tourism-dependent economies, the impact of the September 11 attacks was most evident in Jamaica. The current account deficit as a share of GDP widened from 4.4 percent in 2000 to 7.3 percent in 2001 deteriorating further to 11.0 percent in 2002. Public finances also deteriorated. The overall fiscal deficit increased more than fivefold to 4.9 percent of GDP as revenues fell and there was an upturn in expenditures.

A number of Caribbean SIDS increased their borrowings in the external market, borrowing mainly on commercial terms from private creditors, helping to increase not only the level of debt but the cost of servicing these obligations. Jamaica's external debt service ratio increased from 6 percent in 2000 to 16 percent in 2002 reflecting in part the higher cost of borrowing as well as the fall in export earnings.

Rising Food and Fuel Prices

Almost all Caribbean SIDS are highly import-dependent, particularly on food and on fuel. The steep increases in food and petroleum prices in 2006–2008 therefore represented a significant shock to the economies of these food and fuel importers. The rise in food and fuel costs translated into steep increases in import costs and rapidly deteriorating trade balances. With the exception of commodity-exporting Trinidad and Tobago which benefited from the higher oil export prices, all other Caribbean SIDS saw a widening of their current account balances, which moved from an average 15 percent of GDP in 2005 to 21 percent in 2007 (see figure 3.5).

Some countries, such as The Bahamas and St. Lucia, experienced a doubling of their current account deficit at the onset of the food–fuel

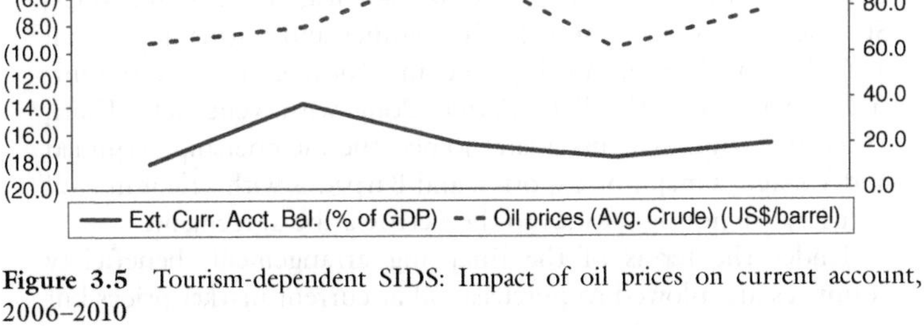

Figure 3.5 Tourism-dependent SIDS: Impact of oil prices on current account, 2006–2010

Source(s): International Monetary Fund (2013c) and World Bank (2013).

crisis. The Bahamas' current account jumped from 8.5 percent of GDP to 17.7 percent, while St. Lucia's increased twofold to 30.4 percent. Deficits increased by around 50 percent in other high net importing Caribbean SIDS, such as Antigua and Barbuda, Dominica, Jamaica, and St. Vincent and the Grenadines. The high oil prices also had a negative impact on travel and tourism, as costs associated with travel increased and demand for the tourism product fell. This further debilitated countries' current accounts.

The large trade imbalances and the financing gaps it created led to an increased need for external financing. As a result, public external debt increased in many countries. Lending was particularly sought from Venezuela under the terms of the 2005 PetroCaribe agreement (Box 3.1). This agreement provided for a share of the oil import costs to be financed on highly concessionary terms. All PetroCaribe members took advantage of this financing to help provide cash flow relief on their external accounts. As a consequence of these borrowings, public debt increased across the board. In Jamaica, some 40 percent of the country's oil bill was financed by PetroCaribe. From an amount equivalent to 1.8 percent of GDP at the end of 2006, debt obligations as a share of GDP grew threefold, rising to 6.9 percent at end 2010.

Box 3.1 PetroCaribe arrangement and the Caribbean

PetroCaribe is an agreement between Venezuela and mainly oil-importing Caribbean states to purchase oil from Venezuela on preferential payment terms. The objective of the PetroCaribe was to alleviate the burden of rising oil prices and the adverse effect on Caribbean economies but also to encourage South–South relations and regional integration. The PetroCaribe agreement came into effect in June 2005 with the signing of the Energy Cooperation Agreement (ECA).

Fourteen countries—Antigua and Barbuda, The Bahamas, Belize, Dominica, Grenada, Guyana, Jamaica, St. Kitts and Nevis, St. Lucia, St. Vincent and the Grenadines and Suriname, as well as Cuba, the Dominican Republic and Nicaragua—were the initial signatories to the PetroCaribe along with Venezuela. Haiti, Honduras, and Guatemala later joined the membership. Trinidad and Tobago, a major oil exporter, and Barbados with a fledgling oil industry are not signatories to the PetroCaribe agreement.

Under the terms of the financing arrangement, beneficiary countries are allowed to purchase oil at current market prices but pay only a percentage of the cost upfront. The percentage to be financed is determined by the oil price, with an increasing share of the oil bill deferred as oil prices rise. The percentage to be financed was determined as follows:

Barrel Price / Percentage Financed

- Greater than or equal to US$15 per barrel / 5
- Greater than or equal to US$20 per barrel / 10
- Greater than or equal to US$22 per barrel / 15
- Greater than or equal to US$24 per barrel / 20
- Greater than or equal to US$30 per barrel / 25
- Greater than or equal to US$40 per barrel / 30
- Greater than or equal to US$50 per barrel / 40
- Greater than or equal to US$100 per barrel / 50

If oil prices are US$40 per barrel or lower, the remaining balance is paid over 17 years including a two-year grace period. Where oil prices exceed US$40 per barrel, the repayment period is extended to 25 years inclusive of a two-year grace period, with an applicable rate of interest of 1 percent per annum. Beneficiary countries are allotted a quota of oil purchases on these concessionary terms, amounting in total to 185,000 barrels of oil per day.

A condition of PetroCaribe is that savings from oil concession are to be used for social and economic development within the beneficiary countries. Since its inception, social projects under PetroCaribe have been in a variety of sectors including health and education, agriculture, housing, environment, sports and road networks while infrastructure projects have included the building of refineries and distribution plants as well as the expansion of power generation plants in a number of member countries.

Source(s): PetroCaribe; www.petrocaribe.org

Increases in the price of imported food and fuel transmitted quickly into domestic price inflation in Caribbean SIDS. Interest rates in many countries rose at the same time that domestic consumption fell, leading to a decline in economic activity. From 5.7 percent in 2006, inflation in Jamaica surged to 16.8 percent, while in the ECCU inflation tripled from 2.2 percent to 6.0 percent. By 2007, Jamaica's Treasury bill rates which were fairly stable at 12 percent had surpassed the 20 percent threshold.

Policy responses among countries were varied but mainly involved measures to stimulate demand and protect the socially vulnerable. Food subsidies were introduced in Jamaica; value-added tax (VAT) on selected food items was reduced or eliminated in Dominica and St. Vincent and the Grenadines; while Barbados scaled up income transfers under social welfare programs. The Bahamas, Dominica, and St. Kitts and Nevis raised wages. The combined deterioration in the fiscal and external accounts in 2006–2007 halted the nascent decline in debt-to-GDP levels that were evident in 2004 through to 2006. By 2008, with the spillover of the food and fuel crisis into the 2008 financial crisis, public debt as share of GDP had begun its ascent (International Monetary Fund 2010).

Global Economic and Financial Crisis of 2008–2009

The 2008 global financial crisis profoundly affected the Caribbean and tourism-dependent SIDS in particular. As the financial crisis plunged the US economy into recession and quickly pulled the major industrial nations into a similar economic slump, tourist arrivals and remittances to the Caribbean fell sharply, leading to GDP declines. Sovereign spreads on emerging market bonds also jumped markedly (International Monetary Fund 2008). These factors both contributed to significant increases in the public debt.

Among all Caribbean SIDS, tourism arrivals dropped by around 5 percent in 2009, the year after the crisis, and among ECCU countries fell even more sharply by 11.5 percent (see figure 3.6 and table 3.12). The downturn in tourist arrivals significantly affected tourism receipts in Caribbean SIDs, which in turn had significant adverse effects on the external and fiscal accounts as well as on employment and growth.

To cushion the impact of the shock to the tourist industry, many Caribbean SIDS relaxed their fiscal stance and implemented measures to stave off anticipated declines in output and employment. Countries, such as Antigua and Barbuda, St. Kitts and Nevis, and St. Vincent and the Grenadines provided short-term tax relief to hotels as an incentive to prevent hotel layoffs and maintain employment levels. Other countries, such as Jamaica and The Bahamas, embarked on aggressive marketing campaigns and heavily discounted travel packages to encourage visitor arrivals (International Monetary Fund 2009).

Many ECCU states provided or increased short-term income transfers to the unemployed to ease the social and economic impact of the tourism fallout. Public expenditures as a share of GDP increased in all Caribbean SIDS with the exception of Barbados, where government spending as a share of GDP fell modestly. Antigua and Barbuda saw expenditures as share of GDP increasing from 27.1 percent in 2008 to 36.9 percent in 2009, while Jamaica saw its ratio rise from 35 percent to almost 40 percent over the period.

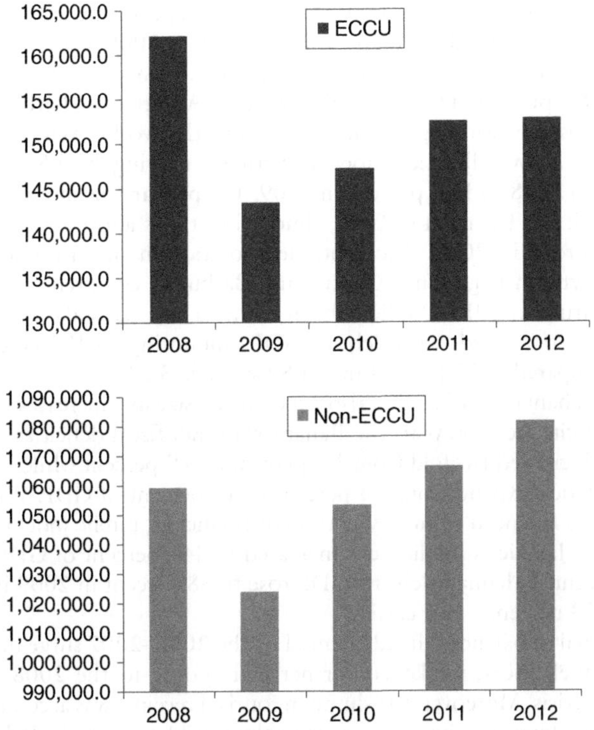

Figure 3.6 Caribbean SIDS: Average tourist arrivals (stay-over) by sub-region, 2008–2012

Table 3.12 ECCU: Tourist stay-over arrivals pre- and post-global financial crisis

ECCU Member States	Arrivals pre-global crisis, 2007	Arrivals on onset of global crisis, 2008	Arrivals one year after global crisis, 2009	Arrivals two years after global crisis, 2010
Antigua and Barbuda	261,802	265,844	234,410	229,943
Dominica	72,098	75,052	69,042	70,618
Grenada	129,088	123,770	109,474	105,419
St. Kitts and Nevis	123,062	127,705	93,081	98,329
St. Lucia	287,518	295,761	278,491	305,937
St. Vincent and the Grenadines	89,532	84,101	75,446	72,478

Source(s): Eastern Caribbean Central Bank (2012) and Caribbean Tourism Organisation (2011).

The fall in tourism receipts and increases in expenditure led to significant weakening of public finances. From 2008 to 2009, fiscal deficits in Caribbean SIDS almost tripled, moving from an average 2.8 percent of GDP to 6.4 percent of GDP. Fiscal balances in Antigua and Barbuda, one of the most tourism-dependent countries in the world, were the most affected. The overall deficit more than tripled, moving from 5.7 percent of GDP in 2008 to 18.2 percent in 2009. The primary balance worsened to a deficit of 11 percent of GDP, almost four times above the 3 percent ratio recorded in 2008. The acute deterioration in the fiscal accounts was catalytic in triggering Antigua and Barbuda's comprehensive debt restructuring in 2010. Public debt in Antigua and Barbuda jumped by more than one-third, with debt-to-GDP amounting to 102 percent in 2009 compared to 77 percent in 2008 (see table 3.13).

The Bahamas and Jamaica also experienced sizeable increases in their fiscal deficits. Year on year, the Bahamas' overall fiscal deficit as a share of GDP increased twofold from 1.8 percent to 3.9 percent while Jamaica saw its deficit expand from 7.4 percent to 11 percent of GDP. With fiscal deficits financed primarily with debt financing, public indebtedness increased. Jamaica's public debt increased to 141 percent of GDP from 126 percent. Bahamas' debt-to-GDP rose to 38 percent in 2009 in contrast to 33 percent a year earlier.

External imbalances already dented by the 2006–2007 surge in world oil and fuel prices, weakened further in response to the 2008 global financial crisis. Moreover, the slump in tourism receipts was accompanied by a sharp downturn in remittances to the Caribbean, particularly from the United States. The external current account deficit for Caribbean

Table 3.13 Caribbean tourism-dependent SIDS: Debt-to-GDP pre- and post-global financial crisis

Caribbean SIDS	Year immediately prior (2007)	Year of financial crisis (2008)	Year immediately after onset (2009)	Year 2 after onset (2010)
Antigua and Barbuda	93.3	76.9	102.0	90.6
Bahamas, The	30.8	32.6	38.1	45.5
Barbados	53.2	56.1	63.9	72.6
Dominica	90.9	64.3	63.3	69.9
Grenada	110.0	83.7	97.1	102.4
Jamaica	114.8	126.2	141.2	143.0
St. Kitts and Nevis	134.0	131.0	148.5	163.9
St. Lucia	55.3	56.2	60.4	65.2
St. Vincent and the Grenadines	55.6	57.0	64.8	66.8

Source(s): International Monetary Fund (2013c).

SIDS slid to an average 17.4 percent of GDP in 2008 from 16.2 percent in 2007, with Barbados suffering the worse decline. Jamaica, the highest recipient among Caribbean SIDS of remittances from abroad, also recorded a steep decline in remittance inflows, by some 15 percent, from 2008 to 2009 (Clarke 2009; Bank of Jamaica 2011). The dual effect of falling tourism earnings and reduced remittances saw Jamaica's external current account deficit widen to 18.4 percent of GDP from 16.9 percent in 2008.

As tourism receipts plummeted, several Caribbean SIDS sought financial assistance from the International Monetary Fund to bridge their external financing gaps (Table 3.14). St. Vincent and the Grenadines was the first country among Caribbean SIDS and in the Caribbean and Latin American region to request a drawing under the IMF's Exogenous Shock Facility (ESF). This facility provides concessional financing to eligible countries needing balance of payments support as a result of sudden and exogenous shocks. St Vincent and the Grenadines received SDR 3.74 million or 45 percent of quota. Dominica and St Lucia also quickly thereafter sought disbursements under the ESF to help absorb the shock to their economy's tourism sector.

Countries, such as Grenada, that were already receiving IMF support also requested additional assistance to augment their finances. Grenada in November 2008, requested a one-year extension and an SDR2.98 million augmentation of its 2006 Poverty Reduction and Growth Facility to help cope with the significantly reduced tourism earnings. Antigua and Barbuda and Jamaica also requested assistance from the International Monetary Fund through Stand-By Arrangements (SBAs) to address their balance of payments problems.

The resort to IMF funding not only added to the countries' debt burden but also changed the composition of the external debt portfolio.

Table 3.14 Caribbean SIDS requesting IMF assistance in response to 2008 global financial crisis

Year	Country	Operation	Amount (SDR millions)
2008	St Vincent and the Grenadines	Exogenous shock facility	3.74
	Grenada	Poverty reduction and growth facility	2.98
2009	Dominica	Exogenous shock facility	3.28
2009	St Lucia	Exogenous shock facility	6.89
2010	Antigua and Barbuda	Stand-by arrangement	
2010	Jamaica	Stand-by arrangement	

Source(s): Various IMF articles for various years.

As the IMF and regional development banks played an increasing role as lender of last resort, the share of debt owed to official bilateral donors declined, to be replaced by an increasing share of obligations to multilateral financial institutions.

Domestic Conditions

Interest Expenditures

High and rising interest costs place an increasing burden on governments' fiscal accounts and, if not accompanied by increases in primary surpluses, lead to further debt accumulation. Most Caribbean SIDS, saw rising domestic interest rates during the second half of the 2000s, particularly in the wake of the food and fuel crisis, when price inflation soared (see table 3.15). Interest expenditures as a share of GDP rose steadily as a consequence of tighter monetary policies implemented to curb inflation and, in the case of countries such as Jamaica, to also maintain exchange rate stability.

The impact of rising interest rates was most pronounced in Jamaica, especially over the period 2007–2009. Jamaica's economic downturn and worsening macro economic conditions, precipitated by the food–fuel–financial crisis, led to reduced investor confidence. This exerted pressure on the exchange rate. To halt a sliding Jamaica dollar, the authorities tightened monetary conditions by raising central bank rates and the cash reserve ratio. The bank rate increased by more than 600 basis points to 21 percent and Treasury bill yields rose steeply. With a high share of floating rate debt benchmarked to Treasury bill yields, the interest burden on domestic debt rose (see figure 3.7). Already high rates on government securities, which exceeded 20 percent in the mid-2000s, spiked above 30 percent in 2009. The government's interest bill jumped from an average 15 percent of GDP in 2006–2007 to over 20 percent in 2008, consuming 70 percent of government revenues. The primary

Table 3.15 Domestically indebted Caribbean SIDS: Three-month treasury bill rates

Countries	2001	2002	2003	2004	2005	2006	2007	2008	2009	2010	2011	2012
Antigua and Barbuda	7.00	7.00	7.00	7.34	7.28	7.29	6.91	6.86	7.07	7.22	6.92	5.00
The Bahamas	2.60	2.00	1.57	0.32	0.35	3.00	3.00	2.70	2.60	2.30	1.30	0.58
Barbados	1.97	1.51	0.64	2.76	6.26	6.56	4.90	4.81	3.44	3.35	3.43	3.61
Dominica	6.40	6.40	6.40	5.20	5.47	6.70	6.71	6.62	6.51	6.35	6.35	4.35
Jamaica	15.70	15.68	19.86	13.90	12.34	11.60	12.51	21.79	15.50	7.21	6.26	6.94
Trinidad and Tobago	6.49	4.39	4.76	4.70	5.01	6.74	7.00	6.94	1.36	0.37	0.28	0.39

Source(s): Bank of Jamaica (2013), Central Bank of Bahamas (2013), and Central Bank of Barbados (2013).

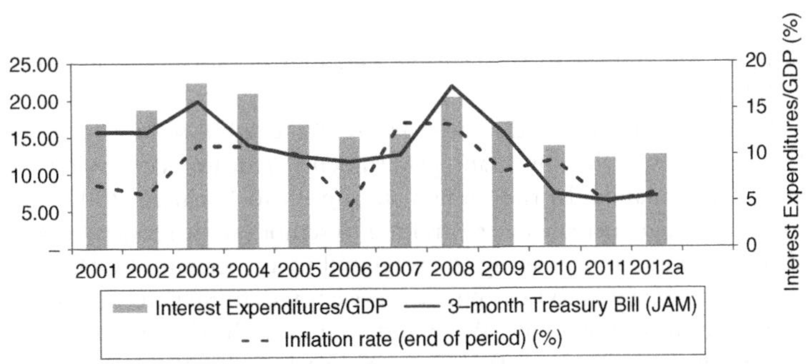

Figure 3.7 Jamaica: Movements in inflation, T-bill rates, and interest expenditure/GDP, 2001–2012

Source(s): Bank of Jamaica (2013) and International Monetary Fund (2013).

surplus shrank from 8.0 percent in 2007 to 5 percent in 2008 and the public debt-to-GDP ratio rose from 115 percent to 128 per percent.

Market-access countries, namely Barbados and Jamaica, saw credit spreads widen after the onset of the 2008 financial crisis, increasing issuance costs in the international capital markets. After averaging 800 basis points for much of 2000–2005, Jamaica's credit spreads widened to 1,100 basis points in 2008. Both Jamaica and Barbados increased their reliance on domestic market financing through the issuance of securities. A consequence was the further pile-up of debt in both countries and an overall increase in their debt burden.

Wage Bill

High fiscal expenditures, particularly on public wages and salaries, play a significant role in explaining the rise in public debt among Caribbean SIDS. Public wages and salaries constitute the largest share of total expenditure among Caribbean SIDS (Amo-Yartey et al. 2012), amounting to more than 40 percent of total current expenditures in countries such as Antigua and Barbuda. In terms of GDP, the wage bill is also high, exceeding the generally accepted threshold of 9 percent. The wage bill in most Caribbean SIDS has exceeded this benchmark and has averaged around 11 percent of GDP in most countries throughout 2000–2012.

In many Caribbean SIDS the government has been the largest employer of labor and has arguably been highly protective of its role as "employer of the last resort" (International Monetary Fund 2005). Fiscal consolidation efforts have therefore been frequently undermined by the authorities' explicit reluctance to retrench public service workers to cut the wage bill. Governments have argued that with a large unskilled or semi-skilled labor force and a small private sector, reducing the size of the public sector is neither socially nor politically viable. Thus in periods of economic downturn, besides the interest bill, the high wage bill has been the major contributor to the sizeable deficits run by Caribbean SIDS, which have been financed by debt.

The high wage bill reflects both the size of the public service and large salary increases, reflecting many governments' repeated concessions to large wage demands. In the immediate run-up to Jamaica's 2002 elections, large wage increases were granted to senior public service workers and parliamentarians, pushing the wage bill to 12 percent of GDP. The intent at the time was to raise public service wages to 80 percent of comparable private sector wages. Wage and salaries in 2002 were increased, in some cases by almost 100 percent. Similarly larger than anticipated increases were awarded to teachers and nurses in the run-up to the 2007 general election, leading to higher expenditures, larger deficits, and rising debt.

Where significant job losses have occurred in the wake of external shocks, many Caribbean SIDS have sought to either maintain employment levels or introduce welfare benefits such as employment insurance or cash transfer programs to minimize social dislocation. Only since the late 2000s and under IMF supported economic programs have some Caribbean SIDS, such as Antigua and Barbuda, and Jamaica, adopted wage bill containment measures through social pacts involving a wage or hiring freeze.

The difficulties in reducing the interest and wage bill have limited the scope for fiscal consolidation through cuts in recurrent expenditure. Revenue-raising measures including the introduction of value-added tax and improved tax administration, as well as one-off inflows from divestments, have been relied on to narrow fiscal deficits. In addition, capital expenditures have often been slashed to bring about targeted reductions in the deficit. Not only has this constrained growth generating activities but it also has limited government's ability to pay down the debt.

Unproductive Public Capital Expenditure

Frequently, the borrowing used to fund public investment projects in Caribbean SIDS has not generated significant streams of revenue. A striking example relates to the large capital expenditures associated with the regional hosting of the 2007 International Cricket Conference (ICC) World Cup of Cricket (see table 3.16). Significant government spending was directed to upgrading or building new stadiums as well as roads, airports, hotels, and marinas funded by a mix of debt and grants. Private investors received huge incentives through tax concessions for hotel-related construction to increase tourism capacity (International Monetary Fund 2007). The large sporting and tourism expenditures were justified on the widely held view of, "if you build it, they will come" (Erikson and Lawrence 2008).

Table 3.16 Capital expenditures/GDP in selected Cricket World Cup host countries, 2004–2009

CWC Host Countries	2004	2005	2006	2007	2008	2009
Antigua and Barbuda	2.5	4.0	8.2	6.4	6.5	2.1
Grenada	8.0	15.0	20.2	13.5	11.7	8.3
Jamaica	1.9	1.9	3.8	4.1	4.0	3.1
St. Kitts and Nevis	6.2	6.7	7.3	7.3	5.8	6.2

Source(s): International Monetary Fund (2004, 2007, 2009).

The anticipated significant legacy effects from the cricket event did not occur. Tourist arrivals fell sharply across the Caribbean including the cricket-hosting nations, as the global financial crisis eroded household incomes in the United States, the Caribbean's main source market. Instead, Caribbean SIDS were left with increased debt levels and underutilized facilities.

Contingent Liabilities and Off-Budget Activities

Contingent liabilities, both explicit and implicit, have been a source of significant fiscal distress for a number of Caribbean SIDS and have led to substantial increases in public debt. Government explicit contingent liabilities have arisen mainly through the issuance of loan guarantees, a practice common across all Caribbean SIDS. Government implicit contingent liabilities have largely arisen from financial sector bailouts. While such events have been fewer, the impact on the fiscal accounts and the public debt has generally been far more serious.

Governments have often favored *explicit* contingent liabilities because, faced with tight budget constraints, they allow the pursuit of public policy objectives while bypassing the scrutiny associated with explicit budget items. Explicit contingent liabilities have often been an effective means of obscuring governments' true budgetary costs, at least in the short term. At the end of 2011, explicit contingent liabilities in the forms of government guarantees exceeded 12 percent in several SIDS including Antigua and Barbuda and Jamaica.

Caribbean SIDS also amassed debt arising from substantial *implicit* contingent liabilities during the 2000s. The public debt of several sugar and banana exporting Caribbean SIDS increased when, with the dismantling of the EU's preferential trade arrangements for these products, the liabilities of loss-making, state-owned sugar companies were assumed by their respective governments.

Implicit contingent liabilities have also increased debt in Jamaica, and more recently, in Trinidad and Tobago and Antigua and Barbuda. In 2001, following the 1996–1997 financial sector collapse, the government of Jamaica assumed financial sector liabilities amounting to 40 percent of GDP on its balance sheet. Jamaica's public debt jumped by 22 percentage points, from 49 percent of GDP in 2001 to 71 percent in that year.

The 2009 failures of Antigua and Barbuda's Stanford Financial Group and Trinidad-based CL Financial Limited affected virtually all Caribbean SIDS except Jamaica. In Trinidad and Tobago, two insurance companies, British American Insurance Company (BAICO) and Caribbean Life Insurance Company (CLICO), collapsed in the wake of poor financial practices and weak regulation. The collapse of BAICO and CLICO led to wholesale runs on deposits in branches across the Caribbean, but particularly in ECCU countries where exposure (15 percent of GDP) was the greatest. A liquidity support fund was established by the governments of the ECCU, Barbados, and Trinidad and Tobago to support the liquidity position of BAICO operations in the ECCU. In Trinidad and Tobago, the cost to the government of the bailout over the period 2009 to early 2013 amounted to US$3.4 billion. Public debt-to-GDP increased by some 13 percentage points.

The 2009 collapse of the Stanford Financial Group amid fraud charges levied by the US Securities and Exchange Commission (SEC) triggered a bank run in Antigua and Barbuda and, ultimately, the failure of the Bank of Antigua, a Stanford-owned bank. The bank collapse led to job losses which in turn adversely affected the fiscal accounts through revenue losses. The provision of funding to Antigua and Barbuda by the Eastern Caribbean Central Bank (ECCB) to assist in the recapitalization of the intervened bank also added to the debt burden. The loan amounting to 3 percent of GDP aggravated Antigua and Barbuda's already unsustainable debt position and no doubt helped catalyze the comprehensive debt restructuring undertaken in 2010.

The increased reliance on public–private partnerships (PPPs) has also been a factor in increased debt levels in Caribbean SIDS, most notably in Barbados. Since the early 2000s, Barbados has used PPPs as the main channel for executing its public investment program. The authorities have implemented several large scale projects including the building of a prison, a desalinization plant, a judicial center, and a major highway. These arrangements, however, were recorded as off-balance sheet items. In 2007, Barbados shifted to budgetary accounting on an accrual rather than cash basis. As a result, the completed PPPs when included on the government balance sheet added some US$150 million to the public debt.

Natural Disasters

The Caribbean is extremely vulnerable to natural disasters, especially hurricanes. Given the small size of their economies, the financial impact of disasters is substantial. As highlighted in table 3.17, hurricanes typically

Table 3.17 Caribbean SIDS: Selected natural disasters, 2000–2012

Country	Year	Event	Cost and Damage
Antigua and Barbuda	2008	Hurricane Omar	Total damage – $54 million. Agricultural loss of $11 million.
	2010	Hurricane Earl	Total damage – $12.6 million. Fatalities – 1.
Dominica	2001	Hurricane Iris	
	2007	Hurricane Dean	Total damage – $162 million. Fatalities – 2.
Grenada	2004	Tropical Storm Earl	Moderate damage, flooding
		Hurricane Ivan	Catastrophic damage. Total damage equivalent to approximately 200 percent of 2004 GDP. Fatalities – 39.
	2005	Hurricane Emily	Significant damage. Total damage – $110.4 million. Fatalities – 1.
Jamaica	2002	Hurricane Lili	Damage to agriculture and roads – approx. S3 million Fatalities – 4.
	2004	Hurricane Ivan	Catastrophic damage. Total damage equivalent to approximately 200 percent of 2004 GDP. Fatalities – 39.
	2005	Hurricane Emily	Total damage – $65 million. Fatalities – 5.
	2007	Hurricane Dean	Severe damage. Total damage – $327 million (10% of GDP). Fatalities – 3.
	2008	Hurricane Gustav	Total damage – $210 million (2% of GDP). Significant damage to road infrastructure. Fatalities – 15.
	2010	Tropical Storm Nicole	Total damage – $238.6 million. Fatalities – 13.
	2012	Hurricane Sandy	Total damage – $100 million. Fatalities – 1.
St. Kitts and Nevis	2008	Hurricane Omar	Total damage – $19 million (3.5% of GDP). Fatalities – 0
St. Lucia	2002	Hurricane Lili	Total damage – $20 million. Fatalities – 4.
	2007	Hurricane Dean	Total damage – $18 million. Agricultural sector – 75% of crops lost. Fatalities – 1.
		Hurricane Tomas	Heavy damage to agricultural sector. Total damage – $588 million. Fatalities – 14.
	2011	Earthquake	
St. Vincent and the Grenadines	2002	Hurricane Lili	Heavy damage to agricultural sector. Total damage – $20 million. Fatalities – 4.
	2010	Hurricane Tomas	Heavy damage to agricultural sector. Total damage – $288 million. Fatalities – 0.

Note: All figures quoted in US dollars.

cause widespread damage to agriculture, the environment, and to infrastructure, causing considerable losses both nominally and in terms of GDP. Given narrow economic bases and heavy reliance on tourism, economic growth and fiscal revenues often decline dramatically. There are no buffers to finance relief, recovery, and reconstruction and therefore there is little option but to finance disaster recovery through borrowing.

For Caribbean SIDS, an added worry is the increase in frequency and intensity of hurricanes. Category 5 strength hurricanes are the most catastrophic of hurricanes. Sustained wind speeds exceed 157 miles per hour. They are noted for the devastation they cause.[6] The 2000s have witnessed the highest number of catastrophic Category 5 hurricanes passing through the Caribbean than any other decade on record. These hurricanes have caused fatalities as well as massive economic losses. Of the recorded eight Category 5 hurricanes that formed in the Atlantic during the period, three hurricanes—Ivan (2004), Emily (2005), and Dean (2007) —were direct or near hits effecting much damage. Hurricane Ivan, a Category 5 hurricane, hit Grenada in 2004 and caused damage amounting to over 200 percent of GDP. Grenada's economy contracted by 3 percent and revenues declined by over 45 percent (see figure 3.8). With large hurricane expenditures and an overall deficit which threatened to deteriorate substantially, Grenada announced a debt moratorium in 2004, as public debt jumped to 121 percent from 102 percent in 2003, and comprehensively restructured its debt in 2005.

Jamaica was similarly devastated by Hurricane Ivan and took large agricultural and infrastructural losses from storms Emily, Dean, Gustav, Nicole, and Sandy between 2005 and 2012 (see figure 3.9). Similarly, hurricane damage incurred by Dominica and St. Lucia in 2008 was quickly

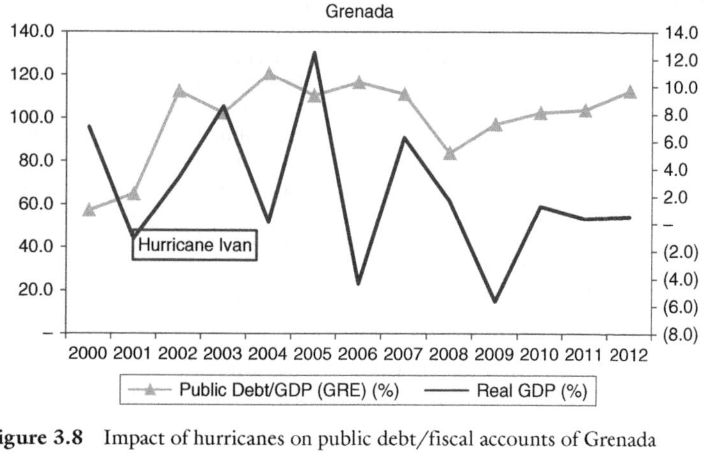

Figure 3.8 Impact of hurricanes on public debt/fiscal accounts of Grenada

Source(s): International Monetary Fund (2012).

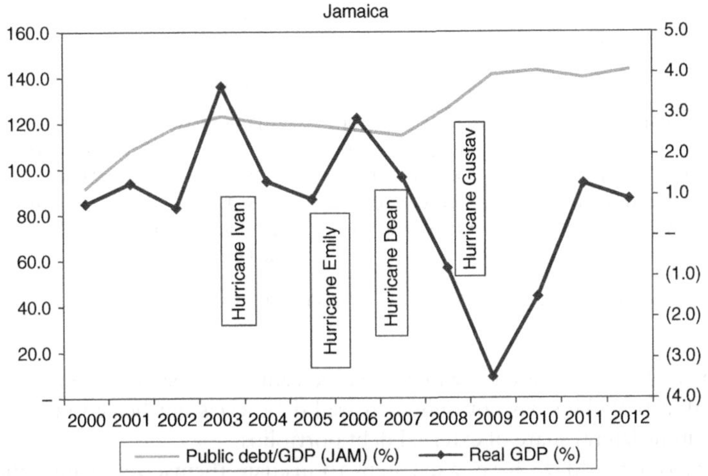

Figure 3.9 Impact of hurricane on public debt/GDP—Jamaica

Source(s): International Monetary Fund (2012).

Table 3.18 IMF emergency assistance to Caribbean SIDS

Country	Year	Event	Amount US$ million	% of quota
Grenada	2003	Hurricane	4.0	25
Grenada	2004	Hurricane	4.4	25
Dominica	2008	Hurricane	3.3	25
St. Kitts and Nevis	2009	Hurricane	3.4	25
St. Lucia	2011	Hurricane	2.4	10

Source(s): International Monetary Fund (2011).

followed by a 7.3 magnitude earthquake in the same year. In all cases, the need for disaster relief expenditures placed already stretched public finances under considerable additional strain. With each event, borrowing both domestically and externally increased and growth slowed, leading to even higher levels of public debt as a share of GDP.

Many Caribbean SIDS sought emergency assistance from the International Monetary Fund and regional development banks to fund recovery and reconstruction programs. As indicated in table 3.18, the IMF Exogenous Shock Facility has been used by four Caribbean SIDS, all from the ECCU, in order to secure relatively quick-disbursing emergency assistance. In 2011, over 50 percent of lending from the Caribbean Development Bank was for emergency reconstruction as a result of natural disasters. This compares to 2007 when only 11 percent of all lending was for emergency assistance and 2008 in the aftermath of Hurricane

Table 3.19 Caribbean catastrophe risk insurance—payouts to Caribbean SIDS (as at June 26, 2012)

Event	Date	Affected Country	Payout (US$)
Earthquake	November 29, 2007	Dominica	528,021
Earthquake	November 29, 2007	St. Lucia	418,976
Tropical Storm Tomas	October 2010	Barbados	8,560,247
Tropical Storm Tomas	October 2010	St Lucia	3,241,613
Tropical Storm Tomas	October 2010	St. Vincent and the Grenadines	1,090,388

Source(s): Caribbean Catastrophe Risk Insurance Facility (2013).

Dean when this share increased to 32 percent (Caribbean Development Bank 2012). This has not only increased debt levels but helped increase the multilateral share of external debt portfolios.

Caribbean SIDS have taken steps to insulate themselves against the financial losses and increased public debt associated with natural disasters by use of the Caribbean Catastrophic Risk Insurance Facility (CCRIF) (Caribbean Catastrophe Risk Insurance Facility 2013). This facility, the first and only one of its kind in the world using parametric insurance, was established in 2005 following the regional devastation caused by Hurricane Ivan and other tropical storms in 2004. All Caribbean SIDS are members of CCRIF.

Unlike traditional insurance payouts which are determined by an assessment and quantification of losses, insurance payouts are triggered upon the occurrence of the event and the specific intensities of the natural disaster.[7] This enables quicker payouts in the event of a natural disaster and thus aids in the delivery of relief without need for financing. However, payments from CCRIF are often only triggered in the most catastrophic events and are still relatively small in relation to funding needs. Thus despite substantial damage from several hurricanes and tropical storms, there have only been five payouts to members over the period 2006–2012 (see table 3.19). As a result, Caribbean SIDS have still had to supplement this funding through borrowing.

Factors Likely to Affect Debt Trends in the Caribbean SIDS in the Near Future

At present, the debt of Caribbean SIDS, with the exception of The Bahamas and Trinidad and Tobago, is unsustainably high, with debt-to-GDP well in excess of the 60 percent benchmark widely accepted as a threshold limit for debt sustainability. ECCU member countries have adopted this benchmark and have committed to achieving it by 2020.

Jamaica, with the second highest debt-to-GDP ratio in the Caribbean and among the top five most heavily indebted countries in the world, has committed to achieving a debt-to-GDP ratio of 96 percent by 2020. Several factors will determine both the debt trajectory and debt composition over time, including the ability of Caribbean SIDS to sustain present fiscal consolidation efforts and to weather the inevitable external shocks associated with global economic developments and natural disasters.

New Borrowing Sources—China and Venezuela

China has become a major donor to the Caribbean as aid from major Western donors, especially the United States and the United Kingdom, has dwindled. China's main condition for aid delivery is the foregoing of diplomatic ties with Taiwan in keeping with its "one China" policy (Girvan 2011). Two Caribbean SIDS, Dominica in 2004 and Grenada shortly thereafter, switched diplomatic ties and benefited from substantial aid inflows. Dominica received US$100 million in grants equivalent to almost one-third of its total GDP. Non-ECCU countries including the Bahamas, Jamaica and Trinidad and Tobago have also benefited from Chinese aid. In 2011, The Bahamas received funding from China to build a new US$3.4 billion hotel resort. In Jamaica, China has financed major infrastructure projects, with loans and grants amounting to over US$600 million.

Chinese aid offers attractive terms with low interest rates (around 1–2 percent) and long repayment periods. For many Caribbean SIDS, particularly market-access countries, this is a welcome relief from high cost market funding. However, as Caribbean debt portfolios include an increasing share of Chinese loans, Caribbean SIDS face increased Yuan Renmimbi exposure which could prove costly over the longer term.

Venezuela has become a major donor to Caribbean SIDS since the 2006 fuel crisis. Assistance has been provided both through the PetroCaribe agreement as well as through the Bolivarian Alliance for the Americas (ALBA). Under ALBA-Caribe, Venezuela has provided development aid to several Caribbean countries for social and economic infrastructure. Loans, both under Petro-Caribe and ALBA, have substantially increased the share of debt owed to Venezuela. When combined, China and Venezuela have overtaken Western donors as the largest contributors of aid to the Caribbean, accounting for more than a 50 percent share of bilateral external debt in ECCU countries such as Dominica and Grenada. In Jamaica, China and Venezuela accounted for over 60 percent of total external debt at the end of 2012 (Government of Jamaica 2013).

Increased Reliance on Multilateral Institutions

The IMF has increasingly assumed the role of lender of last resort as Caribbean SIDS facing unsustainable debt burdens have had little recourse but to seek exceptional support. In the first half of the 2000s, only Dominica received IMF assistance under a Stand-By Agreement. However, in the second half of the decade, four countries—Antigua and Barbuda, Grenada, Jamaica and St Kitts and Nevis—have restructured their debt and requested financial support from the IMF. Lending from other multilateral donors has often been conditional on IMF support, as occurred when US$1.2 billion in support from the Inter-American Development Bank, the World Bank and the European Union (EU) depended on Jamaica securing the US$1.2 billion in financial assistance from the IMF under a Stand-By Arrangement.

Not surprisingly, the share of multilateral debt has increased significantly among the most highly indebted countries. Multilateral debt comprises more than half the total external debt in almost all ECCU countries except Antigua and Barbuda and St. Kitts and Nevis, and even in these countries the share of multilateral debt has increased. Reliance on multilateral funding is likely to continue over the foreseeable future as market access for Caribbean SIDS remains limited given their weak credit ratings and that funding from traditional bilateral sources continues to decline.

Table 3.20 Caribbean small island states: Recent IMF financial arrangements, 2000–2011

Country	Type	Arrangement Date	Expiration Date	Amount Approved (SDRs)	Amount Drawn (SDRs)	Percent of Quota
Antigua and Barbuda	SBA	June 2007	June 2013	67.50	48.88	
Dominica	SBA	Aug 2002	January 2004	2.97	2.97	
	ECF	December 2003	December 2006	7.69	7.69	
Grenada	PRGF	April 2006	April 2010	16.38	6.9	
	ECF	April 2010	April 2013	8.78	2.53	
Jamaica	SBA	February 2010	May 2012	820.5	541.8	
	EFF	December 2009	December 2013	26.40	19.80	
St. Kitts and Nevis	SBA	July 2011	July 2014	52.51	43.10	

Source(s): International Monetary Fund (various reports).

Fiscal Consolidation

The fiscal consolidation efforts of highly indebted Caribbean SIDS will be a critical factor in determining future debt dynamics and whether a sustained reduction in public debt can be achieved. Given the size of fiscal deficits and weak economic activity in most SIDS, fiscal consolidation efforts will have to rely heavily on expenditure reduction and be complemented by revenue-raising measures.

Cutting government spending has proven difficult for many SIDS since the main avenue for reduction—the wage bill and cutting the size of the public sector—has been largely nonnegotiable whether explicitly or implicitly. Many governments have reduced spending by slashing capital expenditures, a measure that undermines efforts to stimulate economic activity. However, reductions in these expenditures have frequently been offset by larger than budgeted wage increases or cash transfers to prevent or minimize the impact of job losses.

The track record of Caribbean SIDS has not been encouraging (Amo-Yartey et al. 2012). In Barbados, fiscal consolidation efforts articulated in a 2010 Medium-Term Fiscal Strategy (MTFS)[8] were derailed in the first year of implementation as weak global economic conditions persisted. While some revenue enhancement measures were implemented, progress on expenditure reduction was slow and while the fiscal deficit narrowed, debt sustainability assessments suggested that it was not sufficient to see a sustained downward path in the debt.

Jamaica's history of sustained fiscal consolidation has been similarly checkered. Under a 2010 Stand-By Arrangement (SBA) with the International Monetary Fund, Jamaica was to build on the success of the recently implemented debt restructuring by cutting expenditures, especially the wage bill, adopting measures to increase revenues and implementing fiscal and debt reforms, including: a new debt management act, a fiscal responsibility bill, pension reforms, and a single treasury account. While there was success in implementing some structural reforms, policy slippages occurred within the first year of the program. The fiscal consolidation effort was undermined by higher-than-anticipated wage increases, lower tax revenues due to the frequent dispensing of tax waivers, and delays in privatization. The 2010 IMF program was abandoned by end 2011 and with the change of political administration a new Fund program has been negotiated in 2012.

Debt sustainability analyses undertaken by the IMF indicate that for highly indebted Caribbean SIDS, debt reduction will not occur unless fiscal consolidation is sustained. Even with a strong commitment to fiscal consolidations, Caribbean SIDS remain at high risk of debt distress. Analyses for Antigua and Barbuda, Barbados, Grenada, Jamaica, and St. Kitts indicate that external shocks such as a natural disaster or upturn in food and fuel prices could place debt on an upward trajectory even with existing consolidation efforts.

Economic Growth

Sustained economic growth is a key factor in achieving continued reductions in debt and a return to debt sustainability. Conversely, weak economic activity or economic decline will maintain public debt at unsustainably high levels. Given current projections,[9] prospects are for weak economic growth, especially among tourism-dependent Caribbean SIDS, over the next few years. In 2013, growth is projected at an average annual 1.2 percent in tourism-dependent SIDS as well as in ECCU member states, while commodity-exporting countries such as Trinidad and Tobago should enjoy two times that growth rate. In 2014, growth is expected to rise to 2 percent, well below that for commodity-exporting countries and for the Latin American and Caribbean region as a whole. With the continued threat of natural disasters, high dependence on tourism exports and food and fuel imports, Caribbean SIDS will have to consider a greater focus on growth-promoting reforms.

Concluding Remarks

High and rising public debt is one of the most pressing economic challenges facing Caribbean SIDS. While public debt as a share of GDP has declined to relatively low levels in African and Asia/Pacific small states over the past decade, among Caribbean SIDS it remains high and rising. Debt to GDP has averaged over 80 percent for all Caribbean SIDS over the period 2001–2010, reaching 87.9 percent at the end of 2012. Among tourism-dependent SIDS, which include all ECCU member states, the situation is worse, with debt to GDP averaging close to 100 percent over the decade and up to 94 percent by the end of 2012.

A number of factors have contributed to the substantial debt accumulation among Caribbean SIDS. Domestic macroeconomic policies, in particular very accommodative fiscal policies, have been a major factor in the persistent increases in debt. Fiscal deficits have averaged over 5 percent of GDP throughout the decade. External shocks have also played a contributory role. Natural disasters have increasingly caused widespread damage, necessitating recovery efforts that have been funded largely by debt. The economic shocks associated with the 2001 US terrorist attacks, the 2007 food–fuel crisis, and the 2008 global financial crisis have deeply affected tourism-dependent economies where both foreign exchange earnings and tax revenue are highly linked to economic behavior in North America and Europe. Caribbean SIDS have sought to maintain employment and stimulate economic growth at the cost of significant increases in debt.

Efforts to contain the growth in public debt have seen periods of loose fiscal policies followed by strong but temporary fiscal adjustment. Caribbean SIDS appear locked into a vicious cycle of debt accumulation and economic stagnation. A growing number of Caribbean SIDS have undertaken debt restructuring operations so as to obtain debt relief and reduce their debt burden. Over the decade Antigua and Barbuda, Dominica, Grenada, Jamaica, and St. Kitts and Nevis have all restructured their debt. The debt profile and experience of Caribbean SIDS suggest that tackling the debt problem will continue to be challenging over the medium term.

Notes

1. For the purpose of this study, the Caribbean is defined as all members of the Caribbean Community (CARICOM) with the exclusion of Haiti.
2. Jamaica is an exception with a current (2012) population of 2.7 million. Nonetheless it is still classified as a small state because it shares all the common features associated with small states.
3. Micro states are defined as states with 2011 populations of less than 200,000.
4. IMF, Eastern Caribbean Currency Union, Article IV Consultation—June 2009.
5. Specifically the Windward Island producers of Dominica, Grenada, St. Lucia and St. Vincent and the Grenadines.
6. For example, Hurricane Katrina, a Category 5 hurricane, was one of the most destructive and costliest disasters ever in United States history.
7. See Caribbean Catastrophic Risk Insurance Facility—Frequently Asked Questions; www.ccrif.org.
8. See Government of Barbados (2010)—Draft Medium-Term Fiscal Strategy.
9. IMF Regional Economic Outlook—Western Hemisphere May 2013.

References

Amo-Yartey, C., M. Narita, G. Peron Nicholls, J. Okwuokei, A. Peter, and T. Turner-Jones. 2012. *The Challenges of Fiscal Consolidation and Debt Reduction in the Caribbean.* Washington, DC: International Monetary Fund.

Bank of Jamaica. 2011. *Remittances to Jamaica—Findings from a Survey of Remittance Recipients.* Kingston: Bank of Jamaica.

———. 2013. *Statistical Digest.* Kingston: Bank of Jamaica.

Barbados Government Information Service. 2010. Retrieved 2013, from BGIS Media. http://gisbarbados.gov.bb/index.php?categoryid=10&p13_sectionid=1&p13_fileid=197.

Browne, C. 2006. *Pacific Island Economies*. Washington, DC: International Monetary Fund.

Caribbean Catastrophe Risk Insurance Facility. 2013. *Frequently Asked Questions—FAQ*. Retrieved May 2013. http://www.ccrig.org.

Caribbean Development Bank. 2012. *Review of CDB's Lending Products—Final Report*. Bridgetown: Caribbean Development Bank.

Clarke, Lavern. 2009. "Remittances Plunge—Flow Dips to Three-Year Low." *Jamaica Gleaner*, August 14. http://jamaica-gleaner.com/gleaner/20090814/business/business1.html.

Erikson, D. P. and J. Lawrence. 2008. *Beyond Tourism. The Future of the Services Industry in the Caribbean*. Ontario: The Center for International Governance Innovation.

Girvan, Norman. 2011. *China's Growing Presence in the Caribbean*. Retrieved May 2013. http://www.normangirvan.info/wp-content/uploads/2011/05/china-in-the-caribbean-2007-2011.pdf.

Government of Barbados. 2010. *Draft Medium-Term Fiscal Strategy*. Bridgetown: Government of Barbados.

Government of Jamaica. 2013. *Medium Term Debt Management Strategy 2013/2014–2015/2016*. Kingston: Ministry of Finance.

International Monetary Fund (IMF). 2005. *Eastern Caribbean Currency Union: Article IV Consultation—Staff Report*. Washington, DC: International Monetary Fund.

———. 2007a. *Regional Economic Outlook: Western Hemisphere—April 2007*. Washington, DC: International Monetary Fund.

———. 2007b. *St. Kitts and Nevis: 2006 Article IV Consultation—Staff Report*. Washington, DC: International Monetary Fund.

———. 2008. *Regional Economic Outlook: Western Hemisphere—April 2008*. Washington, DC: International Monetary Fund.

———. 2009. *Regional Economic Outlook: Western Hemisphere—May 2009*. Washington, DC: International Monetary Fund.

———. 2010. *Regional Economic Outlook: Western Hemisphere—October 2010*. Washington, DC: International Monetary Fund.

———. 2011. *IMF Emergency Assistance: Supporting Recovery from Natural Disasters and Armed Conflicts*. Washington, DC: International Monetary Fund.

———. 2012. *St. Kitts and Nevis—2011 Article IV Consultation and Request for Stand-By Arranagement—Staff Report*. Washington, DC: International Monetary Fund.

———. 2013a. *Caribbean Small States: Challenges of High Debt and Low Growth*. Washington, DC: International Monetary Fund.

———. 2013b. *The Bahamas—2012 Article IV Consultation*. Washington, DC: International Monetary Fund.

———. 2013c. *World Economic Outlook Database—April 2013 Edition*. http://www.imf.org/external/pubs/ft/weo/2013/01/weodata/index.aspx.

Sahay, R. 2005. *Stabilization, Debt and Fiscal Policy in the Caribbean*. Washington, DC: International Monetary Fund.

Standard and Poor's Ratings Services. 2012. *Sovereigns Rating List*. Retrieved June 2013, from http://www.standardandpoors.com/ratings/sovereigns/ratingslist/en/us?sectorName=nul&sub SectorCode=.

UNDP. 2011. *Human Development Index*. Retrieved May 2013, from https://data.undp.org/dataset/Table-1-Human-Development-Index-and-its-components/wxub-qc5k.

UNEP/South Pacific Allied GeoScience Commission (SOPAC). 2012. *Environmental Vulnerability Index*. Retrieved May 2013, from http://www.sopac.org/index.php/environmental-vulnerability-index.

World Bank. 2013. *World Development Indicator Database*. Washington, DC: World Bank. http://data.worldbank.org/indicator/NY.GNP.PCAP.CD.

Chapter 4

The Debt Experience of SIDS in the Pacific

Biman Prasad and Kushneel Prakash

Overview of the Pacific

General Economic Overview of SIDS in the Pacific

The Small Island Developing States (SIDS) in the Pacific region have experienced low economic growth and high volatility in the last few decades. The Pacific SIDS include the island countries of The Cook Islands, Federated States of Micronesia, Fiji, Kiribati, Marshall Islands, Nauru, Niue, Palau, Papua New Guinea (PNG), Samoa, Solomon Islands, Timor-Leste, Tonga, Tuvalu, and Vanuatu. These countries are sometimes collectively referred to as Pacific Island Countries (PICs). Many of the PICs share some common and emerging issues despite the presence of significant diversity in land tenure and property rights, natural resources, and population. Growth and development are being hindered by similar constraints such as limited economies of scale, strong reliance on fluctuating tourism and remittances earnings, vulnerability to natural disasters of all kinds, the impact of external economic shocks, and a narrow production and export base. The Cook Islands, Fiji, Samoa, Palau, and Vanuatu in particular depend heavily on tourism earnings as a major foreign exchange earner while countries such as Kiribati, Samoa, Tonga, and Tuvalu depend heavily on remittances and foreign aid to support their incomes (see table 4.1). Mineral reserve deposits are the primary earners for Papua New Guinea (by far the largest island in the region) and Timor Leste. The rest of the PICs including the Federated States of Micronesia, the Marshall Islands, Nauru, and Niue depend heavily on external assistance and foreign aid from countries such as Australia, New Zealand, and the United States to support income earned from small domestic economic activities.

Table 4.1 Pacific SIDS: Major economic activities

Cook Islands	Fruit processing, tourism, finance, copra, citrus fruits, clothing, coffee, fish, pearls and pearl shells, mining, handicrafts
Federated States of Micronesia	Tourism, construction, fish processing, craft items (shells, wood, pearls), garments, banana, black pepper.
Fiji	Sugar, tourism, copra, gold, silver, clothing, timber, fish processing, cottage industries.
Kiribati	Fishing, handicrafts, copra.
Marshall Islands	Copra, fish, tourism, craft items (shell, wood, pearls), coconut oil.
Nauru	Phosphate mining, financial services, coconut products.
Niue	Taro, cassava, yams, bananas, copra, passion fruit, limes, vanilla, noni, tourism and fisheries.
Palau	Tourism, craft items (shell, wood, pearl), commercial fishing, agriculture.
Papua New Guinea	Copra crushing, palm oil processing, plywood production, woodchip production, mining of gold, silver, and copper, crude oil production, construction, tourism, timber, coffee, cocoa, seafood.
Samoa	Fishing, tourism, timber, food processing, coconut oil and cream, copra, beer.
Solomon Islands	Timber, fish, palm oil, cocoa, copra.
Timor-Leste	Coffee, marble, oil, sandalwood, oil, gas.
Tonga	Tourism, fishing, squash, fish, vanilla, root crops, coconut oil.
Tuvalu	Fishing, tourism, copra, stamps/coins.
Vanuatu	Fishing, offshore financial services, tourism, food and fish freezing, wood processing, meat canning, coconuts, cocoa, coffee.

Source(s): SOPAC (2009).

The 15 Pacific SIDS introduced earlier are distinct and unique in their own ways. These countries are also sometimes divided into three different sub-regions—namely Melanesia, Micronesia, and Polynesia. The Melanesian subregion extends from the western end of the Pacific Ocean to the Arafura Sea and comprises Fiji, Papua New Guinea, Solomon Islands, and Vanuatu. The Micronesian subregion comprises small islands on the western side of the Pacific Ocean including the Federated States of Micronesia, Kiribati, the Marshall Islands, Nauru, and Palau. Scattered over the central and southern regions of the Pacific Ocean are the Polynesian counterparts comprising Niue, Samoa, Timor Leste, Tonga, Tuvalu, and the Cook Islands.

Papua New Guinea has more than two-thirds of the total population of the Pacific SIDS. Timor-Leste also has a population in excess of 1 million while the remaining countries taken together have a population of less than a million. The third most populous country in the region, Fiji, is one of the economic leaders and is an economic hub of the Pacific. Niue with the land area of 2,400 square kilometers is the smallest country in the region and has a population of only around 1,400 people. Around 90–95 percent of the Niuean people live in New Zealand with which they share dual citizenship. By way of introduction, table 4.2 presents numerous indicators that highlight key characteristics of the respective countries.

In what follows, the analysis and tables used for the debt indicators are based purely on data availability and only the countries that have the relevant data sets have been included in the analysis. The countries for which there is poor data availability include the Cook Islands, Nauru, and Niue. However, it should be noted that these countries are very small and rely heavily on development assistance for their budgetary support.

Although most of the Pacific SIDS have been traditionally classified as lower middle-income countries, five countries in the region are now included in the upper middle-income classification with incomes ranging from US$4,086–US$12,615 (see table 4.3). Fiji, the Marshall Islands and Tonga were recently (July 2013) reclassified by the World Bank as upper-middle-income countries. There are no PICs currently classified as either high-income or low-income countries as the per capita income among these countries ranges from a low of US$1,130 in Solomon Islands to a high of US$9,830 in Palau.

The PICs country policy and institutional assessment (CPIA) indicator for 2007 to 2011 revealed that most of the countries have weak policies and institutions in place (see table 4.2). The indicator revealed that with the exception of Papua New Guinea, Vanuatu (both ranked in the medium category), and Samoa which has the strongest policies and institutions in place in the region, the other island countries are plagued with weaknesses in economic management, structural policies, public sector management, and institutions.

In terms of socioeconomic development, as measured by the United Nations Human Development Index (HDI), the performance of Pacific SIDS varies considerably (see table 4.4), ranging from a high of 0.791 in Palau to a low of 0.466 in Papua New Guinea. Only four Pacific SIDS exceed the World average HDI (0.694) and the East Asia and Pacific region average HDI (0.683) as at 2012. Additionally, only Fiji, Palau, Samoa, and Tonga exceed the average for all the Small Island Developing States of 0.648.

The Pacific Island developing states have experienced declining economic growth since the 1990s. Those most affected include Fiji, the Marshall Islands, Solomon Islands, Tuvalu, and Vanuatu (see table 4.5). This is while many other island economies have experienced fluctuating

Table 4.2 Pacific SIDS: Brief overview of selected economic and social indicators

SIDS in the Pacific	Income Classification	IMF CPIA Indicator 2007–2011	Risk of Debt Distress by IMF/WB	Environmental Vulnerability index-status	Economic Vulnerability index	World Bank lending category	Population in millions (2012)	GNI per capita, PPP (current international $)	HDI for 2012
Cook Islands	n.a	n.a	n.a	extremely vulnerable	n.a	n.a	0.011	n.a	n.a
Federated States of Micronesia	Lower middle	n.a	Low	extremely vulnerable	n.a	IDA	0.103	3310	0.645
Fiji	Upper middle	n.a	n.a	highly vulnerable	44.9	IBRD	0.875	4200	0.702
Kiribati	Lower middle	Weak	High	extremely vulnerable	82	IDA	0.101	2260	0.629
Marshall Islands	Upper middle	Weak	Low	highly vulnerable	n.a	IDA	0.053	4140	n.a
Nauru	n.a	n.a	n.a	extremely vulnerable	n.a	n.a	0.009	n.a	n.a
Niue	n.a	n.a	n.a	vulnerable	n.a	n.a	0.001	n.a	n.a
Palau	Upper middle	n.a	Moderate	highly vulnerable	n.a	IBRD	0.021	9830	0.791
Papua New Guinea	Lower middle	Medium	Low	at risk	38.3	Blend	7.167	1790	0.466
Samoa	Lower middle	Strong	Moderate	highly vulnerable	51.1	IDA	0.189	3220	0.702
Solomon Islands	Lower middle	Weak	Moderate	vulnerable	55.2	IDA	0.550	1130	0.530
Timor-Leste	Lower middle	Weak	Low	n.a	53.3	Blend	1.210	3670	0.576
Tonga	Upper middle	Weak	High	extremely vulnerable	59.6	IDA	0.105	4240	0.710
Tuvalu	Upper middle	Weak	High	extremely vulnerable	63.9	IDA	0.010	6070	n.a
Vanuatu	Lower middle	Medium	Low	vulnerable	46.8	IDA	0.247	3080	0.626

Notes: n.a – represent that data is not available for these countries relevant indicator.

Source(s): Data extracted from World Bank (2013a, 2013b), International Monetary Fund (2013c) and United Nations Development Programme (2012).

Table 4.3 Pacific SIDS: 2012 GNI per capita

Country income Classification	Country	2012 GNI per capita, PPP (current international $)
High-income countries (US$12,616 or more)	None	–
Upper-middle-income countries (US$4,086-$12,615)	Palau	9,830
	Tuvalu	6,070
	Tonga	4,240
	Fiji	4,200
	Marshall Islands	4,140
Lower-middle-income countries (US$1,036-$4,085)	Timor-Leste	3,670
	Federated States of Micronesia	3,310
	Samoa	3,220
	Vanuatu	3,080
	Kiribati	2,260
	Papua New Guinea	1,790
	Solomon Islands	1,130
Low-income countries (less than US$1,035)	None	–

Source(s): World Bank (2013b).

Table 4.4 Pacific SIDS: 2012 Human Development Index

Human Development Index	Country	HDI in 2012
Between 0.700–0.799	Palau	0.791
	Tonga	0.710
	Fiji	0.702
	Samoa	0.702
Between 0.600–0.699	Federated States of Micronesia	0.645
	Kiribati	0.629
	Vanuatu	0.626
Between 0.500–0.599	Timor-Leste	0.576
	Solomon Islands	0.530
Between 0.400–0.499	Papua New Guinea	0.466
	Small Island Developing States	0.648
	East Asia and Pacific region	0.683
	World	0.694

Source(s): United Nations Development Programme (2013).

economic growth with even slight improvements in the last two decades. On a positive note, all the island economies in the Pacific SIDS have recorded positive growth rates for 2011 and 2012. There are therefore signs of slow recovery from the 2008–2009 global financial crisis. With the exception of Papua New Guinea, Timor-Leste, and Vanuatu, all the Pacific SIDS experienced a period where there were significant declines in income especially in the mid-2000s. This is, however, now slowly improving. Papua New Guinea, Solomon Islands, Timor-Leste, and Vanuatu are the only countries which have performed better than the Pacific SIDS average in the last two decades. The remaining economies have underperformed while Palau's economy actually contracted by an average of around 0.1 percent since 1990. Besides Palau, the other economies have achieved economic growth of a meager 1 to 2 percent since 1990.

Overview of Public Debt of SIDS in the Pacific

The Pacific SIDS have low levels of public debt despite being relatively small open economies. Though the level of public debt is low, there is still concern regarding the sustainability of debt given the low rates of economic growth recorded (see table 4.5). As at the end of 2012, the Pacific SIDS had a wide range of public debt levels with external debt to GDP ratios being well below 60 percent in some cases. The exception to this rule is the Republic of Marshall Islands with public debt at 60 percent of GDP (see figure 4.1). Overall, the debt levels of these countries averaged around 30 percent of GDP in 2012 with Samoa, Fiji, and Tuvalu along with the Marshall Islands being the countries in the region that were more indebted.

Table 4.5 Pacific SIDS: Real GDP growth rates, 1990–2012

Pacific SIDS	1990–1995	1996–2000	2001–2005	2006	2007	2008	2009	2010	2011	2012
Federates States of Micronesia	5.0	−0.1	0.6	−0.2	−2.1	−2.6	1.0	2.5	2.1	1.4
Fiji	3.2	2.2	2.4	1.9	−0.9	1.0	−1.3	−0.2	2.0	2.2
Kiribati	1.4	6.0	1.2	−0.4	2.1	−2.4	−2.3	1.4	1.8	2.5
Marshall Islands	5.0	−3.2	2.1	1.9	3.0	−1.9	−1.3	5.2	5.0	1.9
Palau	−0.1	1.9	3.5	−1.5	0.6	−5.3	−10.6	−0.8	6.5	5.3
Papua New Guinea	6.9	−0.1	1.6	2.6	7.2	6.7	5.5	8.0	9.0	8.0
Samoa	0.2	4.0	5.0	2.0	1.8	4.3	−5.1	0.4	2.0	1.2
Solomon Islands	8.0	−2.6	1.2	6.9	0.7	7.3	−1.2	7.0	9.0	3.9
Timor-Leste	n.a	13.7	0.1	−3.1	11.6	14.6	12.8	9.5	10.8	8.6
Tonga	2.8	2.0	2.5	−1.6	−4.1	1.2	2.9	2.7	4.9	0.8
Tuvalu	3.2	3.4	0.2	2.6	6.0	6.9	-1.7	−3.0	1.2	1.2
Vanuatu	4.7	3.6	1.0	8.5	5.2	6.5	3.3	1.6	1.4	2.3
Pacific SIDS Average	3.7	2.6	1.8	1.6	3.4	3.0	0.2	2.9	4.7	3.3

Source(s): World Bank (2013a).

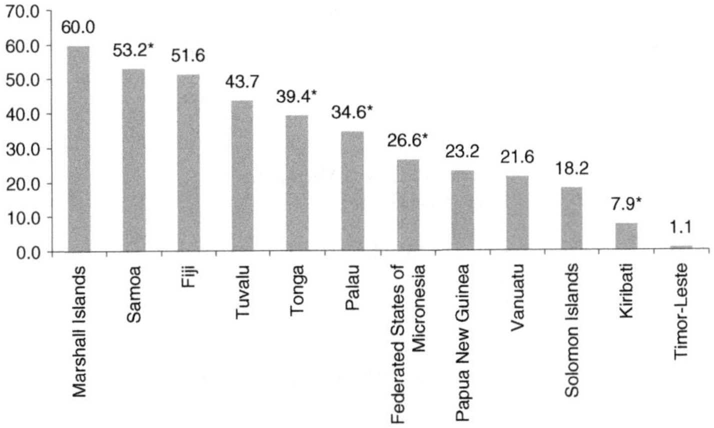

Figure 4.1 Overview of public debt levels in Pacific SIDS in 2012 (% of GDP)

Table 4.6 Classification of Pacific SIDS indebtedness as at end 2012, % of GDP

Low Debt (less than 30%)	Federated States of Micronesia Kiribati Papua New Guinea Vanuatu Solomon Islands Timor-Leste
Low to Medium Debt (0% to 60%)	Fiji Palau Samoa Tonga Tuvalu
High Debt (between 60% to 90%)	Marshall Islands
Very High Debt (Exceeding 90%)	*None*

Source(s): Data extracted from the IMF (2013c) and World Bank (2013a).

The public debt to GDP ratios of below 60 percent shared by most Pacific SIDS classifies them as having low to medium debt (see table 4.6). More specifically, half of the Pacific SIDS in this analysis are classified as having low debt to GDP ratios. The largest country, Papua New Guinea, had a debt to GDP ratio of 23.2 percent in 2012. Timor-Leste had the lowest public debt to GDP ratio of all the Pacific SIDS at only 1.1 percent of GDP. Fiji, Palau, Samoa, Tonga, and Tuvalu have debt ratios ranging between 30 to 60 percent of GDP

Table 4.7 Classification of Pacific SIDS by net external position

Net Creditor	Net Debtor	HIPC
Timor Leste	Fiji	*None*
	Kiribati	
	Marshall Islands	
	Federated States of Micronesia	
	Papua New Guinea	
	Samoa	
	Solomon Islands	
	Tonga	
	Vanuatu	
	Tuvalu	

Source(s): Data extracted from the IMF World Economic Outlook (2013c).

Table 4.8 Credit ratings of Fiji and PNG, Standard and Poor's ratings

Pacific SIDS	Rating	Rated as
Fiji	B	Stable
Papua New Guinea	B+	Stable
Other Pacific SIDS	Not reported	

Source(s): Data extracted from tradingeconomics.com.

and are classified as moderately indebted Pacific SIDS. None of the Pacific SIDS in the analysis has very high public debt levels exceeding 90 percent of GDP.

The Pacific SIDS are mainly classified as net debtors by the IMF country classification, which examines the net external position of countries. This is because, except for Timor-Leste (classified as a net creditor), the rest of the Pacific SIDS' current account balance accumulation for the last three to four decades has generally been negative (see table 4.7). This also means that the value of foreign-owned assets in many Pacific SIDS exceeds the value of their assets owned abroad. Although this is not a direct measure of the indebtedness of a country, it does indicate the reliance of the Pacific SIDS on external funding. Nonetheless, based solely on public debt ratios, none of these island countries can be classified among the Heavily Indebted.

Given the fact that many Pacific SIDS are relatively small with a comparatively narrow resource base, they do not have easy access to international capital markets. Additionally, the major rating agencies do not provide sovereign credit ratings for most of the Pacific SIDS. Of the major agencies, only Standard and Poor's provides a credit rating and only for Fiji and PNG. Both Fiji and PNG's recent credit ratings of B and B+ respectively presents them as being stable economies (see table 4.8).

The Debt Profile of the Pacific SIDS

The Pacific SIDS are some of the least indebted countries in the world, with the majority of their loans coming from external sources and donor agencies, typically carrying a significant amount of concessions. The debt of the countries across the Pacific region varies both in amounts and as percentages of national income. The average total debt among the Pacific SIDS has been gradually declining over the last ten years. It has declined from an average of 38.4 percent in 2000 to 31.8 percent in 2012 (see figure 4.2). This is evidenced by the fact that all the Pacific SIDS except Fiji and Palau had lower debt levels in 2012 than in 2000 (see figure 4.3). Notwithstanding this, some island countries such as the Federated States of Micronesia, Samoa, and Tonga experienced an increase in debt levels in the late 2000s due to increased borrowing for rehabilitation works after natural disasters, and later to cushion the impact of the global economic and financial crisis of 2008–2009.

The debt levels for countries such as the Federated States of Micronesia, Kiribati, Timor-Leste, and Vanuatu have always remained below the average for the Pacific SIDS. In the case of Palau, the debt level exceeded the average level in 2012 despite being below the average in prior years. Papua New Guinea, on the other hand, has seen debt levels decline dramatically, to below the Pacific average in 2012. The remaining Pacific

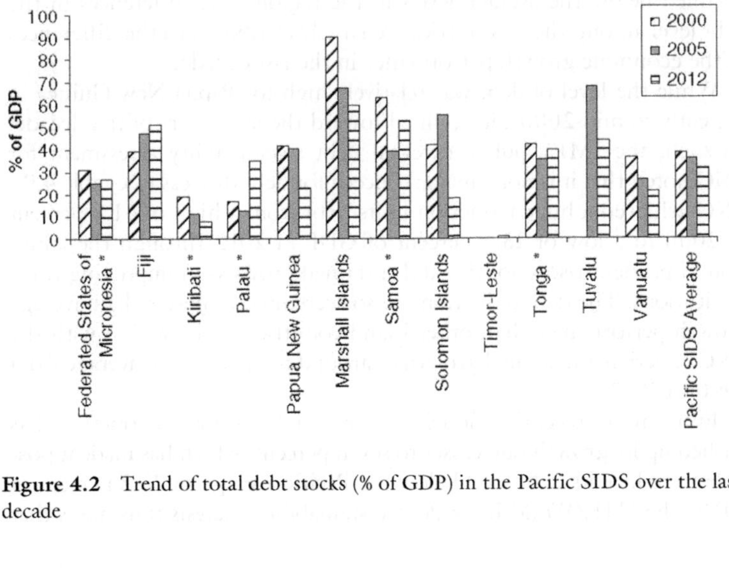

Figure 4.2 Trend of total debt stocks (% of GDP) in the Pacific SIDS over the last decade

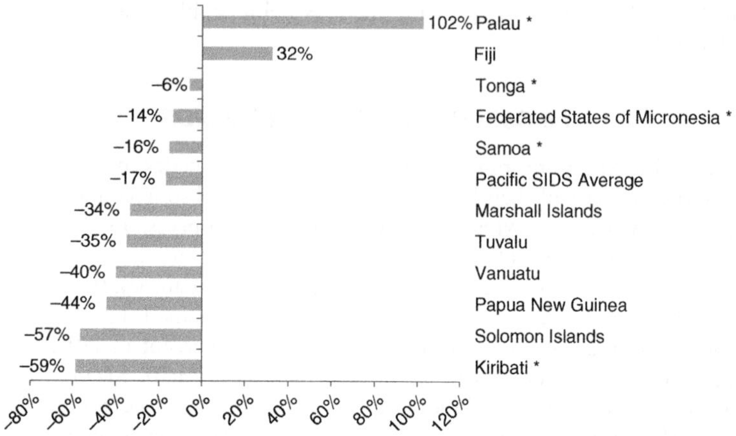

Figure 4.3 Percentage change in debt to GDP ratio of the Pacific SIDS between 2000 and 2012

island countries have experienced debt levels in excess of the Pacific average since 2000.

With the exception of Fiji, the Melanesian countries in the Pacific SIDS including Papua New Guinea, Solomon Islands, and Vanuatu have experienced a gradual decline in debt levels. Only Vanuatu's debt level remained below the Pacific average between 2000 and 2012. More recently, debt levels in Papua New Guinea and Solomon Islands have also declined below the average levels in the region. The differences in the debt level among these countries are largely attributed to the differences in the economic growth performance in the last decade.

While the level of debt was relatively high for Papua New Guinea in the early to mid-2000s, it declined toward the latter part of the decade. As such, the IMF's public external debt sustainability assessment for PNG noted that it is not a major concern for debt distress (see table 4.9). PNG achieved a huge reduction in its debt from a high of 41.8 percent in 2000 to a low of 23.2 percent of GDP in 2012 through the adoption of prudent fiscal policies and sustained progress in improving fiscal institutions. The debt reduction is also accredited to better-than-average growth performance than other island countries in the region with the PNG experiencing annual growth of around 5.5 percent on average from 1990 to 2012.

Even more recently, flourishing mineral resource extraction has pushed up its growth rate closer to seven percent, which has made it possible to reduce its debt level below the Pacific average to 23.2 percent of GDP. The IMF/World Bank debt sustainability analysis thus notes that

Table 4.9 Level of debt distress by IMF/World Bank analysis

High	Kiribati
	Tonga
	Tuvalu
Moderate	Palau
	Samoa
	Solomon Islands
Low	Federated States of Micronesia
	Papua New Guinea
	Marshall Islands
	Timor-Leste
	Vanuatu

Source(s): Data extracted from the IMF article 4 consultation reports of various countries.

PNG has low debt distress, indicating that unstable debt dynamics in PNG shall only occur if the prices of gas, copper, and gold collapse and remain low for at least a decade (IMF 2012a).

The second largest economy in the region, Fiji, experienced a gradual rise in public debt from a low of 39.1 percent of GDP in 2000 to 51.6 percent of GDP in 2012. The growth rate in Fiji has been volatile with sharp swings attributable to political instability. The country has experienced no fewer than four military coups since 1987, which has resulted in sluggish economic growth, averaging slightly more than 1 percent in the last two decades. These uncertainties have pushed its debt levels to one of the highest in the region as the government resorted to borrowing to meet expenditure needs. The IMF noted in its recent assessment that the global financial crisis can also be partly blamed for increasing Fiji's public debt as the country borrowed to cushion the impact of global shocks (IMF 2012b). The report further added that aside from global external shocks the economy also faces downside risks due to domestic political instability and structural weaknesses. The IMF has predicted that it will be hard to improve Fiji's high public debt situation within the next five years as structural reforms are vital to boost potential economic growth (IMF 2012b). However, studies by Narayan and Narayan (2003) and Gounder and Narayan (2007) found that Fiji's budget deficit is actually sustainable in the long run. They argue that borrowing should be done in line with growth prospects as that would determine to a large extent the ability of the government to repay debt in due time. It will therefore be important to maintain effective debt management procedures to prevent the rapid buildup of public debt.

The Solomon Islands is one of the better performing economies in the region. Although it experienced some negative growth rates in the late 1990s and then in 2009, it has on average performed much better than the other Pacific SIDS. On average, the economy has grown by around 5 percent between 1990 and 2012, reaching its peak in 2007 at 10.7 percent. As a result of its impressive growth record, the public debt level of the country has been falling. From a peak of 55 percent of GDP in 2005, public debt declined to only 18 percent of GDP in 2012.

Solomon Islands is the only country in the Pacific region that has undertaken debt restructuring from the Honiara Club Agreement (HCA). The improvement is also attributable to stronger than anticipated revenue from logging and from strong improvements in revenue collections. However, based on the IMF/World Bank debt sustainability assessment, the country's debt profile still remains at moderate risk as it is believed to be sensitive to shocks to non-debt creating flows (IMF 2011a). The ongoing negotiations on the review of the Honiara Club agreement framework, which currently only allows the country to receive grants and not loans, could also possibly see the facilitation of external loans on concessional terms. However, even if the concessional borrowing does go ahead, the moderate risk of debt distress is highly unlikely to decrease as this would require continued efforts to build fiscal discipline and implement reforms that promote broad based economic growth (IMF 2011a).

Vanuatu has also experienced declining debt levels since 2000. The decline in debt is largely attributed to steady positive economic growth since the 1990s. In fact, it is the only country in the region which has recorded net positive economic growth in the last two decades (see table 4.5). As such the IMF/World Bank debt sustainability assessment states that the risk of debt distress is relatively low in Vanuatu and indicates that the public debt level is low and should remain manageable in the near future (IMF 2013b). Over the last decade, the debt to GDP ratio for Vanuatu has remained below the Pacific SIDS average and has been continuously on a decline. It has experienced a significant decline in debt, by around 40 percent, between 2000 and 2012. However, given Vanuatu's environmental vulnerability (see table 4.2), new non-debt sources of revenue are important to supplement aid and remittance flows in order to sustain a high level of infrastructure development without raising the debt to GDP ratio.

Samoa, Timor-Leste, Tonga, and Tuvalu have varying debt ratios. These differences are largely due to differences in growth rates achieved by these countries, with Timor-Leste noted to be the best performer among all the Pacific SIDS in the last two decades. All the other Polynesian countries have, however, experienced growth rates below the Pacific SIDS average.

The Samoan economy experienced a net decline in debt levels from 63 percent of GDP in 2000 to 39.3 percent of GDP in 2005. However, in the wake of the global financial crisis and the post 2009 tsunami reconstruction, debt rose again to 53.2 percent in 2012. After recording a negative economic growth rate of 5.1 percent in 2009 due to the adverse impact of the tsunami and from the slowdown in the global economy, the economy has managed to recover somewhat and experienced an average GDP growth of around 1.5 percent between 2010 and 2012. The economic recovery was supported by multilateral donors such as the Asian Development Bank (ADB) and the World Bank along with bilateral partners, in particular Australia, China, Japan, and New Zealand. Samoa has also benefitted from concessional loans which have made its repayment commitments much less costly. However, the recent debt sustainability assessment by IMF and World Bank indicates that Samoa's debt distress risk has risen from low to moderate levels (see table 4.9) calling for more urgent prudent actions to ensure medium and long-term debt sustainability (IMF 2012c).

As noted previously, one of the strongest economic performers is Timor-Leste. Timor-Leste's economic growth has averaged more than 8 percent in the last two decades with its economic growth rate reaching a high of 14.6 percent in 2008. This growth is mainly based on the export of crude oil. Remarkably, Timor Leste did not have any debt until 2012. In 2012, the government approved concessional loans of US$43 million from the World Bank, the ADB and some bilateral donors to finance fiscal deficits expected from increased planned capital expenditure for infrastructure development. This resulted in the creation of public debt recorded at 1.1 percent of GDP in 2012. The debt sustainability assessment indicates that Timor-Leste has a low risk of debt distress despite increased expected borrowing in the future. This is largely credited to the fact that the better than average economic growth for the country can compensate for any economic shortfalls.

Like Samoa, Tonga has recorded an average GDP growth rate of around 1.5 percent over the last two decades. Public debt levels in the economy since 2000 have been higher than the Pacific SIDS average of around 39 percent of GDP. Although the economy experienced a reduction in debt from 42 percent of GDP in 2000 to 35.4 percent in 2005, it thereafter increased to 39.4 percent in 2012. Even though the economy is classified in the ranks of upper-middle-income countries, its high vulnerability to natural disasters and economic shocks has kept its risk of external debt distress relatively high (see table 4.9). The major concern at the moment is Tonga's debt repayment commitment to China's EXIM Bank on two separate loans for reconstruction and road development. Servicing these loans may pose significant fiscal challenges unless reforms in tax collections and spending are carried out. The relatively high debt levels and lower than average economic growth for Tonga calls for reforms to reduce its debt distress and ensure debt sustainability in the longer term.

Tuvalu appears to be in a situation similar to Tonga since it has recorded higher debt levels than the Pacific average since 2005. Nevertheless, the total debt in Tuvalu declined from 67.7 percent to 43.7 percent of GDP in 2012. The country, classified in the ranks of upper-middle-income countries, has halted official borrowing since 2008. However, the main issue still remains with current existing outstanding debts and new loans from unofficial sources. Reforms in financial and revenue collection institutions along with improved economic performance could go a long way to lessen the problem of debt distress in the economy. Tuvalu is extremely vulnerable to environmental shocks and unstable macroeconomic conditions. Its level of debt distress is high due to a lower than average economic growth rate. The IMF/World Bank debt sustainability assessment also points out that the high debt distress can only be contained if donors continue to provide assistance only in the form of grants to meet the country's development needs (IMF 2012e).

The Micronesian countries (the Federated States of Micronesia, Kiribati, the Marshall Islands, and Palau) have all experienced economic growth lower than the Pacific SIDS average since the 1990s. Their combined growth rate in the last two decades has averaged slightly less than 1 percent, far below the Pacific SIDS average of 2.7. As such, the debt to GDP ratios in Palau and the Marshall Islands are some of the highest in the region. It is, however, also important to note that all the Micronesian countries except for Palau have achieved reductions in public debt levels between 2000 and 2012. Specifically, the Federated States of Micronesia has experienced a gradual decline in debt levels in the last decade. Its debt level has remained below the Pacific SIDS average debt level since 2000 and as a result the IMF/World Bank debt sustainability assessment considers it as having low debt distress. Being classified as a lower middle-income country, the country attracts loans on concessional terms from institutions such as the ADB which has kept its debt levels relatively low for a long time.

Although Kiribati has experienced economic volatility since 1990 its debt to GDP ratio has remained well below the Pacific SIDS average in the last decade. Its debt to GDP ratio has continuously declined from 19.4 percent of GDP in 2000 to a low of 7.9 percent of GDP in 2012. However, given its weak CPIA indicator and being extremely vulnerable to adverse climate change impacts and economic shocks, the IMF/World Bank debt sustainability assessment indicates that Kiribati may be in high debt distress. The assessment further adds that securing grant financing and containing fiscal deficits are key to ensuring external debt sustainability (IMF 2011b). The uncertainty about the global economic environment and the impact of climate change continues to add pressure to the fiscal position as the country tries to undertake infrastructure development. Given its low GDP growth rate, most of the financing needs are met through loans from development partners. Therefore, the effort to reduce its debt distress will require additional concessional borrowing and grant financing along with reforms in its fiscal institutions.

The highest debt level in the Pacific SIDS belongs to the Marshall Islands at 60 percent of GDP in 2012. Debt levels in the Marshall Islands have always exceeded the Pacific SIDS average with the highest being recorded in the year 2000 at around 90 percent of GDP. The country is classified as an upper-middle-income country but still qualifies for loans under the World Bank's IDA (International Development Association) category on concessional terms with relatively long grace periods. Therefore, the majority of the borrowing in the country is on concessionary terms. Given this, the IMF/World Bank's debt sustainability assessment indicates that the high debt level in the country is not currently a concern.

Palau is the second most indebted country among the Micronesian group with its debt level at around 35 percent of GDP in 2012, having risen from only 17 percent in 2000. This is largely attributed to the economy contracting by as much as 10.6 percent in 2009 due to the impact of the global financial crisis. The Palau economy has been virtually stagnant over the last two decades. However, being classified as upper-middle-income country, it falls under the World Bank's International Bank for Reconstruction and Development (IBRD)[1] lending category which means that it does not qualify for concessional loans. According to the IMF/World Bank debt sustainability assessment, the high vulnerability to adverse climate change impacts and poor economic growth suggests that the country is in moderate level of debt distress.

External versus Domestic Debt

Public debt in the Pacific SIDS is dominated by loans from external sources (see figure 4.4). Given the small economic resource base for many of the island countries in the Pacific (particularly those classified as lower middle-income countries), many find it easier to borrow externally on concessional terms. There are, however, some exceptions like Fiji and Papua New Guinea that tend to borrow more from domestic sources. These countries are relatively well endowed with natural resources and have a broader economic base. Fiji and Papua New Guinea have domestic debt to GDP ratios of 40 percent and 14 percent, respectively. All the other countries in the region have huge external debt components with little to no domestic debt.

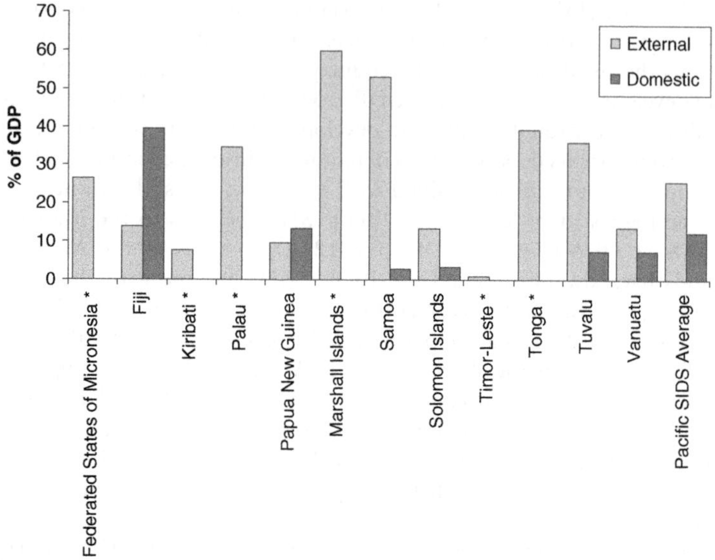

Figure 4.4 External and domestic component of public debt in 2012 (as a % of GDP)

Debt Servicing Needs of the Pacific SIDS

Since many of the Pacific SIDS qualify for loans on concessional terms, debt servicing as a percentage of foreign earnings in the period 2000 to 2005 remained low at around 9.5 percent. It marginally increased to an average 10.2 percent between 2006 and 2012 (see table 4.10). The debt service ratio has remained near or below 10 percent for most of the Pacific SIDS over the review period.

Countries such as the Federated States of Micronesia, the Marshall Islands, and Samoa which qualify for loans on concessional terms have achieved declining debt service ratios in recent years. Fiji, which borrowed a total of FJ$300 million from the international market in 2006, repaid its debt in 2011 and borrowed an additional FJ$500 million in 2011. This subsequently resulted in an increase in the debt service requirement for Fiji in 2011 to 16.6 percent of its foreign earnings. Palau, on the other hand is classified as an upper-middle-income country and does not qualify for concessional loans. This has resulted in it having to repay debts averaging close to 60 percent of the country's foreign earnings in the last four years. Tonga has also ended up repaying relatively high amounts of debt in the recent years partly due to their loans from the EXIM Bank of China. Timor-Leste which did not have any loans up

Table 4.10 Total debt service (% of exports of goods, services, and income in Pacific SIDS)

Pacific SIDS	2000–2005	2006–2008	2009	2010	2011	2012
Federated States of Micronesia	18.8	5.8	6.6	6.6	6.9	6.6
Fiji	2.3	1.9	2.2	1.4	16.6	1.2
Kiribati	7.5	6.6	7.9	3.0	2.6	2.7
Palau	1.6	17.5	69.9	62.4	48.3	52.5
Papua New Guinea	8.1	5.7	1.8	1.4	1.3	1.5
Marshall Islands	39.9	40.2	24.8	22.2	18.4	17.1
Samoa	8.4	4.8	5.1	2.4	3.8	4.5
Solomon Islands	7.5	4.7	3.6	3.0	1.9	1.4
Timor-Leste	0.0	0.0	0.0	0.0	0.0	0.0
Tonga	9.4	9.9	5.6	9.8	6.9	7.4
Tuvalu	n.a	17.2	15.4	11.2	10.7	14.5
Vanuatu	1.6	1.3	1.6	1.4	1.4	1.4
Pacific SIDS Average	9.5	9.6	12.0	10.4	9.9	9.2

Source(s): Data extracted from various countries IMF Article 4 consultation reports for various years.

until 2012, does not have any debt servicing burden from their loans as yet. Even their US$43 million borrowing in 2012 is not expected to be much of a problem to service as most of it is on concessionary terms from the World Bank and the ADB. While reducing its debt levels, Papua New Guinea has also managed to gradually reduce its debt service payments in the last decade along with Fiji, Solomon Islands, and Vanuatu, which have ended up paying less than 2 percent of foreign income flows as debt servicing in 2012. The rest of the Pacific SIDS including the Federated States of Micronesia, Samoa, and Tonga average less than 10 percent of debt servicing as a percentage of exports of goods, services, and income flows in 2012.

Current Factors Affecting Debt Trends in the Pacific SIDS

Due to the high reliance of the Pacific SIDS on tourism and remittances, these countries were greatly impacted by the slowdown of the global economy after the 2008–2009 global financial crisis. Although not having a direct impact, the financial crisis has had adverse indirect effects on the inflow of tourists, sale of exports, and reduced remittance flows. These indirect effects have cumulatively contributed to a lowering of income for both households and national governments. This has been partly blamed for causing the lower post-crisis economic growth rates across the region.

Impact of Global Economic and Financial Crisis

The global financial crisis of 2008–2009 had a significant effect on most of the developed countries to which the Pacific islands are linked. The impact on trading partner countries such as Japan and the United States had their adverse impact through reduced tourist numbers and remittances for the Pacific SIDS. However, other key partner economies, Australia and New Zealand remained resilient and so the Pacific was somewhat cushioned. Additionally, China continued to provide development assistance and concessional loans to the Pacific countries. Despite this cushion, the global downturn did put added pressure on Pacific countries to meet current and capital expenditure needs.

Tourism is one of the major income sources for most of the islands. Tourist arrivals to these countries declined in the midst of the financial crisis (see table 4.11). In the years prior to the crisis, tourist arrivals in the Pacific SIDS increased by an annual average of 17 percent. However, since the crisis tourist numbers increased only marginally. Fiji, Kiribati, the Marshall Islands, Palau, Timor-Leste, Tonga, Tuvalu, and Vanuatu recorded reduced numbers of tourist arrivals. The only exceptions to this trend are Papua New Guinea and Solomon Islands. Both countries have managed to marginally increase tourist arrivals, even during the crisis, due to strong tourism marketing and the diversion of tourists from Fiji due to an unstable economic environment created by a political coup in 2006. Nonetheless, visitor arrivals are slowly increasing in almost all the island countries as economies start to recover from the crisis. Fiji in particular devalued its currency by 20 percent in 2009 to boost tourism numbers and export revenues. As a result, tourist arrivals in Fiji increased by 16.6 percent in 2010.

Table 4.11 Pacific SIDS: Changes in tourist numbers, 2006–2011

Pacific SIDS	2006–2007	2007–2008	2008–2009	2009–2010	2010–2011
Fiji	-1.6	8.3	-7.4	16.6	6.8
Kiribati	6.8	-17.0	0.0	20.5	12.8
Marshall Islands	16.7	-14.3	-16.7	0.0	0.0
Palau	7.3	-10.2	-8.9	19.4	26.7
Papua New Guinea	33.3	9.6	10.5	15.9	13.0
Samoa	6.4	0.9	3.4	0.0	-0.8
Solomon Islands	27.3	14.3	12.5	16.7	9.5
Timor-Leste	57.1	63.6	22.2	-9.1	27.5
Tonga	17.9	6.5	4.1	-11.8	2.2
Tuvalu	0.0	54.5	-5.9	6.3	-29.4
Vanuatu	19.1	12.3	8.8	-2.0	-3.1
Pacific SIDS Average	17.3	11.7	2.1	6.6	5.9

Source(s): World Bank, World Development Indicator Database (2013a)

Table 4.12 Pacific SIDS: Changes in personal remittances, 2006–2011

Selected Pacific SIDS	2006–2007	2007–2008	2008–2009	2009–2010	2010–2011
Fiji	-18.4	-24.1	45.4	-7.8	-24.5
Papua New Guinea	52.0	-23.3	-34.7	-38.7	137.5
Samoa	2.2	-1.4	25.8	-12.6	2.8
Solomon Islands	-15.7	-24.9	55.2	-40.4	-21.7
Tonga	26.1	-19.5	-17.0	-13.5	-14.0
Vanuatu	-7.8	39.1	28.6	-10.9	64.8
Selected Average	11.2	-13.0	5.4	-12.6	-4.9

Source(s): World Bank World Development Indicators Database (2013a).

Remittances which are another major source of revenue for many of the Pacific SIDS were also adversely affected by the global slowdown from 2008 to 2012. In the years prior to the crisis, remittances had been gradually increasing. The most affected economies after the crisis have been Papua New Guinea and Solomon Islands but the former subsequently recovered in terms of increased remittance flows (see table 4.12). The subsequent rise in income and stability in the world economy had a favorable impact on the island economies.

Both the reduction in tourist arrivals and the fall off in remittance receipts during the global economic crisis caused contractions in national incomes and have led to increased borrowing in the Pacific SIDS during the late 2000s. Debt in the crisis year increased significantly for most of the Pacific countries (see figure 4.5). On average, the debt of the Pacific SIDS increased by around 7 percent from 2008 to 2009. In particular, countries such as the Federated States of Micronesia, Fiji, Palau, Samoa, Tonga, and Tuvalu increased borrowing in order to cushion the impact of the economic and financial crisis.

Vulnerability to Climate Change

The Pacific SIDS are highly vulnerable to the adverse effects of climate change. The threat of rising sea levels, extreme weather patterns, and a regular occurrence of natural disasters put an extra burden on the fiscal balances of these countries. Adaptation to climate change impacts require a significant investment. Although the international commitment to provide climate change funding would go a long way in helping Pacific countries, significant additional resources would be required for infrastructure related expenditure and some of these will have to come from loans.

The growth rates and economic progress of the Pacific islands have been adversely affected by damage from natural disasters such as hurricanes, cyclones, floods, droughts, earthquakes, and even tsunamis. These

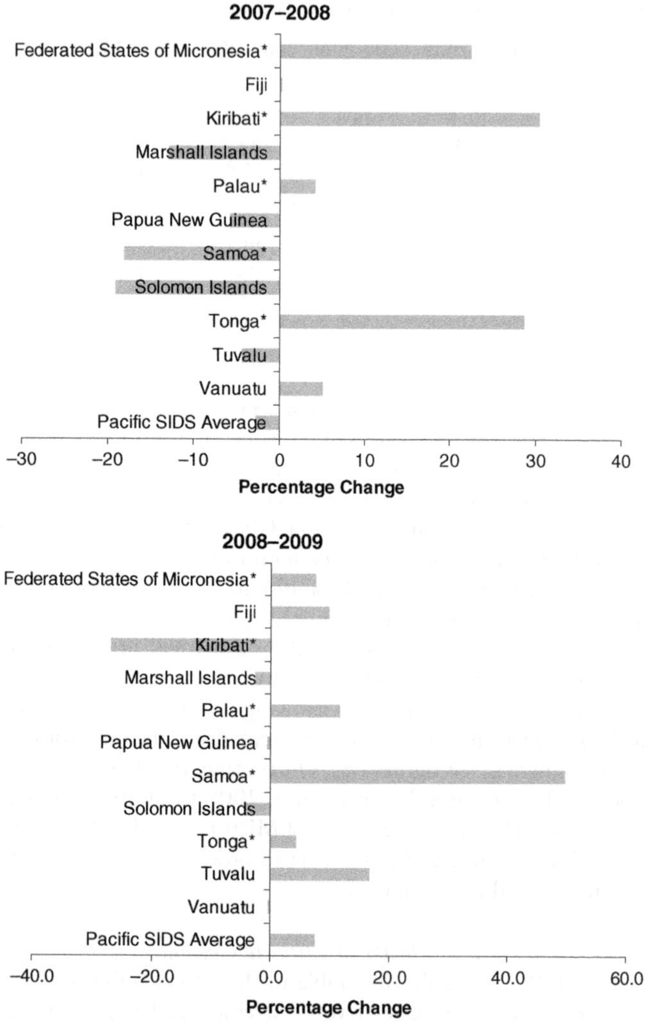

Figure 4.5 Pacific SIDS: Pre and post-global financial crisis changes in debt levels

economies are highly vulnerable to changes in environmental conditions (see table 4.13). Countries such as Fiji, Kiribati, Samoa, and Tonga are constantly faced with risks of natural disasters which can severely impact their livelihood. On the other hand, due to its large size and geographical location, Papua New Guinea remains at risk but is not as exposed to the extent of others in the region.

Table 4.13 Pacific SIDS: Environmental vulnerability status

Extremely Vulnerable	Highly Vulnerable	Vulnerable	At Risk	Resilient
Cook Islands	Fiji	Niue	Papua New Guinea	None
Federates States of Micronesia	Marshall Islands	Solomon Islands		
Kiribati	Palau	Vanuatu		
Nauru	Samoa			
Tonga				
Tuvalu				

Source(s): International Monetary Fund (IMF) Article 4 consultation reports, various country reports.

Since most of the Pacific Island countries are low lying, they are among the most vulnerable to the effects of climate change and rising sea levels in the world. In 2009, a tsunami devastated resources and infrastructure in Samoa and Tonga. Also, the effects of floods in 2009, cyclone Tomas in 2010, and cyclone Evan in 2012 damaged Fiji's economy. To rebuild after these natural disasters, governments have had to resort to borrowing in order to reshape the economy. Urgent efforts such as planting mangroves and constructing sea walls to adapt to climate change are essential to provide for some climatic resiliency. This requires significant funding, much of which may have to be borrowed.

Eligibility Criteria

Debt management in the Pacific SIDS depends to a large extent on the economic characteristics of the country, structure of the domestic financial markets, capacity of the economy to absorb shocks, and a sound record of international borrowing. Pacific countries primarily fall in the category of lower middle-income economies with the exception of Fiji, the Marshall Islands, Palau, Tonga, and Tuvalu, which are upper middle-income economies (World Bank 2013b). With relatively low levels of GDP per capita, many Pacific SIDS qualify for deep concessional loans under the IDA[2] fund established by World Bank to assist poor countries. Fiji and Palau, however, do not qualify for these concessional loans since their GDP per capita are higher than the IDA loan qualification threshold. They do qualify for funds under the IBRD fund. Papua New Guinea, based on both its low per capita income and financial creditworthiness, qualifies for loans from both development funds. These differences in eligibility criteria for Pacific SIDS have greatly influenced the total debt levels and debt servicing requirements of these island countries.

As an interesting case study, Solomon Islands is the only country in the Pacific SIDS which has undergone debt restructuring. Since the country's debt restructuring in 2005 with debt relief from the Honiara Club 2005–2007, it has achieved a reduction in debt levels (Commonwealth Secretariat 2012). The Honiara Club agreement framework allows the country to only receive grants and not loans. This has been fundamental in reducing debt levels in the country.

Factors Likely to Affect Debt Trends in the Pacific SIDS in the Near Future

The Pacific SIDS are some of the least indebted countries in the world largely due to loans received on concessional terms and the availability of significant amounts of grants for expenditure works. The island states, largely classified as low-income countries, are highly vulnerable to natural calamities and external shocks. The high vulnerability of Pacific SIDS to environmental and economic shocks (see table 4.13) is very likely to play a significant role in determining future debt trends in the Pacific.

Increasing pressure from climate change, natural disasters, and changing global economic conditions will continue to impact the Pacific SIDS. The ability of these islands to adapt to climate change and the efforts to mitigate its effects remains a policy challenge. The UNESCAP report (UNESCAP 2013) highlights that adapting to climate change can be turned into an opportunity to create a new approach to development based on sustainability. The ability of Pacific Island countries to strengthen economic ties among each other could also go a long way in benefitting the region as a whole. Undoubtedly remittances and tourism earnings will continue to contribute significantly to growth performances in these economies in the near future as well. In particular, the declining trends of remittances and tourist arrivals will have to be reversed to achieve better growth and keep debt levels at a manageable level.

The IMF has in the past played a part in the Pacific SIDS, as elsewhere, as a lender of last resort, and may in the future be expected to lend support and provide relief from the debt burden of some of these countries. Currently, there are no Pacific SIDS under any financial arrangements with the IMF, though there have been some such arrangements in the past. Papua New Guinea engaged in a stand-by agreement with IMF three times between 1991 and 2001 borrowing amounts between SDR 35 million to SDR 85 million in 2000 (IMF 2012a). The support from IMF played a significant role in reducing the debt burden during a time of need in PNG. Additionally, the IMF also stepped in to support Samoa on two instances in the early 1980s approving stand-by funds of around SDR 3 million (IMF 2012c). More recently, Solomon Islands was also supported by two separate stand-by facilities between 2010 to 2012 with funds totaling around SDR 18 million (IMF 2011a). This played a significant role in debt restructuring in Solomon Islands and has helped bring down the overall debt levels.

Furthermore, Pacific SIDS will continue to require the support of development partners for infrastructure development as many challenges remain in relation to roads, the water supply, airports and seaports, and much else. In the long-term, Pacific SIDS like other developing, small islands, will need to achieve sustained growth rates to satisfy the rising expectations of their people and to stave off debt-burdened stagnation. Strategies must also be put in place to broaden their revenue base as they cannot be assured of long-term donor assistance in the future at the same levels enjoyed in the past. As such, sustained economic growth is vital for reduced dependence on grants and aid along with the capacity to meet debt servicing requirements in the future.

Concluding Remarks

Increasing levels of debt without also raising the capacity to meet debt servicing in the future is a concern in many developing countries. The Pacific SIDS, by contrast, traditionally had relatively little debt, with further declining debt levels since the turn of the century. The debt level in some Pacific SIDS, however, increased during the global economic downturn in the late 2000s. Given the fact that many Pacific SIDS do not enjoy sustained and high economic growth rates; rising debt levels to any extent could present a serious policy challenge to fiscal sustainability.

The ability of Pacific countries to secure grant financing instead of loans is key to ensuring their debt sustainability. Pacific economies that are highly vulnerable to environmental changes and that are largely classified as lower middle-income countries are able to secure loans on highly concessional terms. This has been fundamental in reducing their debt burden and creating much needed fiscal space in the economy.

Debt restructuring in countries with high debt levels similar to the one undertaken in the Solomon Islands could also prove to be influential. Countries such as Fiji, Palau, the Marshall Islands, Samoa, Tonga, Tuvalu, and Vanuatu which have relatively high debt levels or debt servicing needs could potentially benefit from debt restructuring. Such restructuring, however, has to be supplemented by improved political, financial, and commercial institutions, which would help to increase investments while ensuring maximum spillover benefits to other sectors of the economy and enhancing the ability to meet debt obligations as they fall due.

Ultimately, the key challenge to policy makers is to find strategic ways to boost economic growth potential in Pacific countries, as this will lessen the debt burden. Also important in this respect is the way borrowed funds are used. Since most of the island countries in the Pacific rely on exports of agricultural products and proceeds from the service sector for growth, loans diverted toward strengthening of their export capacity and service delivery could be well justified. A multi-year budget framework as proposed by the IMF (IMF 2011a), would help to prioritize the areas that are essential in promoting broad based growth in addition to providing fiscal discipline and while ensuring debt sustainability for the long term.

Notes

1. IBRD (International Bank for Reconstruction and Development) loans are provided by World Bank but are non-concessional in nature.
2. International Development Association (IDA) loans are given by World Bank to qualified borrowing countries on huge concessional terms.

References

Commonwealth Secretariat. 2012. "The Debt Challenges of Commonwealth Small Vulnerable Economies: Trends and Policy Options." Paper Prepared for the Commonwealth Meeting on Debt Challenges of Commonwealth Small Vulnerable Economies (CSVEs), Tokyo, Japan 9. London: Commonwealth Secretariat.

Gounder, N. and P. K. Narayan. 2007. "An Empirical Investigation of the Relationship Between Government Revenue and Expenditure: The Case of Fiji Islands." *International Journal of Social Economics* 34(3): 147–158.

International Monetary Fund (IMF). 2011a. *Solomon Islands: 2011 Article IV Consultation*. Washington, DC: International Monetary Fund. www.imf.org/external/pubs/ft/scr/2011/cr11359.pdf.

———. 2011b. *Kiribati: 2011 Article IV Consultation*. Washington, DC: International Monetary Fund. http://www.imf.org/external/pubs/ft/scr/2011/cr11113.pdf.

———. 2012a. *Papua New Guinea: 2012 Article IV Consultation*. Washington, DC: International Monetary Fund. www.imf.org/external/pubs/ft/scr/2012/cr12126.pdf.

———. 2012b. *Fiji: 2011 Article IV Consultation*. Washington, DC: International Monetary Fund. www.imf.org/external/pubs/ft/scr/2012/cr1244.pdf.

———. 2012c. *Samoa: 2012 Article IV Consultation*. Washington, DC: International Monetary Fund. www.imf.org/external/pubs/ft/scr/2012/cr12250.pdf.

———. 2012d. *Tonga: 2012 Article IV Consultation*. Washington, DC: International Monetary Fund.

———. 2012e. *Tuvalu: 2012 Article IV Consultation*. Washington, DC: International Monetary Fund. www.imf.org/external/pubs/ft/scr/2012/cr12166.pdf. http://www.imf.org/external/pubs/ft/scr/2012/cr12259.pdf

———. 2013a. *Vanuatu: 2013 Article IV Consultation*. Washington, DC: International Monetary Fund. www.imf.org/external/pubs/ft/scr/2013/cr13169.pdf.

———. 2013b. *World Economic Outlook Database*. Washington, DC: International Monetary Fund. http://www.imf.org/external/pubs/ft/weo/2013/02/weodata/index.aspx.

Narayan, P. K. and S. Narayan. 2003. "The Sustainability of Fiji's Budget Deficit: An Econometric Analysis." *Pacific Economic Bulletin* 18(2): 77–90.

SOPAC. 2009. *Relationship between Natural Disasters and Poverty: A Fiji Case Study*, Suva: Secretariat of the Pacific Community. http://www.unisdr.org/files/11851_11851R25PovertyAFijiCaseStudylowres.pdf.

United Nations Development Programme (UNDP). 2012. *Human Development Index*. New York: United Nations Development Programme. https://data.undp.org/dataset/Table-1-Human-Development-Index-and-its-components/wxub-qc5k.

United Nations, Economic and Social Commission for Asia and the Pacific (ESCAP). 2013. *Forward-looking Macroeconomic Policies for Inclusive and Sustainable Development*. www.unescap.org/pdd/publications/survey2013/download/Economic-and-Social-Survey-of-Asia-and-the-Pacific-2013.pdf.

World Bank. 2013a. *World Development Indicators Database [online]*. Washington, DC: World Bank. http://databank.worldbank.org/data/views/variableSelection/selectvariables.aspx?source=world-development-indicators.

———. 2013b. *Data: Country and Lending Groups*, Washington, DC: World Bank. www.data.worldbank.org/about/country-classifications/country-and-lending-groups.

Chapter 5

The Debt Experience of SIDS in the Atlantic, Indian Ocean, Mediterranean, and South China Sea

*Kaymara Barrett, Altricia Dawson, and Sidonia McKenzie**

Overview of the AIMS SIDS

The Atlantic, the Indian Ocean, as well as the Mediterranean and South China Seas (AIMS), are home to a number of Small Island Developing States (SIDS), including the Kingdom of Bahrain, Comoros, Maldives, Mauritius, Seychelles, Cabo Verde, Guinea Bissau, São Tomé and Principe, Cyprus, and Singapore.[1] These countries have population sizes ranging from 88,000 (Seychelles) to over 5,300,000 (Singapore) (see table 5.1). Guinea Bissau, the largest country in the region, occupies 28,120 square kilometers of land followed by Cyprus with 9,241 square kilometers. The majority of the islands in the region, however, occupy less than 1,000 square kilometers of land.[2]

Despite falling into the AIMS regional grouping, the geographic dispersion of these countries contributes to a unique level of diversity not experienced by the Caribbean and Pacific SIDS. Comoros, Maldives, Mauritius, and Seychelles are located in the Indian Ocean, while Cabo Verde, Guinea Bissau, and Sao Tome are found in the Atlantic off the coast of Africa. Cyprus and Singapore are located in the Mediterranean and the South China Seas, respectively, with the former being a European Union member state.

Diversity is also seen in the dispersion of wealth between these countries. The regional average per capita income of US$13,492, which is bolstered by the economic success of developed countries such as Bahrain and Singapore, hides the disparity in wealth between the countries found in the four regional subgroupings (see table 5.2). The African states attained average per capita income of US$1,871 in 2012, with the

Table 5.1 AIMS SIDS: Selected social, economic, and indebtedness indicators

	Number of Islands	Population in Thousands (2012)	Public Debt	
			World Ranking*	Percentage of GDP (2012)
Atlantic Ocean				
Guinea-Bissau**		1664	118	52.8
São Tomé	2	188	144	77.2
Cape Verde	10	494	157	85.9
Indian Ocean				
Bahrain	33	1318	60	33.6
Comoros	4	718	93	42.5
Mauritius		1291	111	50.3
Maldives	1192	338	150	81.3
Seychelles	116	88	152	82.7
Mediterranean Sea				
Cyprus	1	1129	154	85.8
South China Sea				
Singapore	63	5312	168	111.0
AIMS SIDS		**12,540**		**70.3**

Notes: *The ranking was compiled using the available debt data for 177 countries (International Monetary Fund 2013e).
**Guinea-Bissau consists of a mainland and islands.

Source(s): Data extracted from World Bank (2013b), International Monetary Fund (2013e), UN-OHRLLS (2013).

region's poorest country, Guinea Bissau, earning only US$521 in per capita income. SIDS in the Indian Ocean are mainly upper middle-income countries earning on average US$10,174 per capita. This subregion however includes Comoros, a low-income country with per capita income of US$859 in 2012. The region is home to not just some of the poorest countries in the world, but also one of the wealthiest, Singapore. In 2012, Singapore's per capita income was US$52,052, which is 28 times the income of the African SIDS. The Mediterranean state of Cyprus is also classified as a high-income country, with per capita income of US$26,389.

Using the United Nations Human Development Index (HDI) as an indicator of development, there has been a slow but steady increase in regional living standards, improving by 7 percent between 2000 and 2012.[4] In 2009, the average level of development of AIMS SIDS surpassed that of all SIDS. This was owing to the high level of development in countries such as Bahrain, Cyprus, Seychelles, and Singapore, which are among the most developed in the world. In 2012, half of the AIMS small states boasted high to very high levels of development while two countries achieved medium levels of development. Three countries in the region struggled to develop resulting in a low development classification.

Table 5.2 AIMS SIDS: Income and level of development

Country	Income Classification	GDP per capita (Current US$, 2012)	Economic Vulnerability	Level of Development: HDI (2012) Index	Level of Development: HDI (2012) Classification	Risk of Debt Distress
Singapore	High	52,052	30.6	0.895	Very High	n.a
Cyprus	High	26,389	n.a.	0.848	Very High	n.a
Indian Ocean SIDS (Average)	Upper middle	10,173	43.4	0.691	High	
Bahrain	High	23,555	37.4	0.796	High	n.a
Seychelles	Upper middle	11,240	44.5	0.806	Very High	n.a
Mauritius	Upper middle	8,850	29.8	0.737	High	n.a
Maldives	Upper middle	6,363	55.2	0.688	Medium	High
Comoros	Low	859	49.9	0.429	Low	High
Atlantic SIDS (Average)	Lower middle	1871	47.3	0.492	Low	
Guinea Bissau	Low	521	60.5	0.364	Low	Moderate
São Tomé and Principe	Lower middle	1,406	46.1	0.525	Low	High
Cape Verde	Lower middle	3,687	35.2	0.586	Medium	No rating*
AIMS SIDS (Average)		13,492		0.667		

Note: *An explicit rating for Cape Verde was not expressed in the list of Low Income Countries Debt Sustainability Analysis for Poverty Reduction and Growth Trust (PRGT)-Eligible Countries as of May 1, 2014. This is the case for some countries for which IDA does not require a rating for operational purposes. The Debt Sustainability Analysis for Cape Verde dated November 18, 2009 categorized Cape Verde as Low Distress.

Source(s): World Bank (2013b), International Monetary Fund (2013e, 2014b), United Nations Development Programme (2013a).

Human development was notably aligned to subregional groupings as the states in the China and Mediterranean Seas had very high levels of development. This is followed by a generally high level of development in the Indian Ocean with the exception of Comoros and the Maldives, which have low and medium levels of development, respectively. The countries in the African region have the lowest level of human development in the AIMS, as they continue to grapple with improving economic growth and development after debt relief.

The main industries in the region are tourism and agriculture with the exception of the petroleum producer Bahrain and Singapore's specialization in electronics (see table 5.3). The region's main trading partner is Europe, which makes these small states highly vulnerable to shocks to the European economy. Saudi Arabia, India, and Pakistan are the other major trading partners with notable levels of intra-regional trade between the Maldives and Singapore. Like many small states, the AIMS countries are heavily dependent on fuel and food imports due to the

Table 5.3 AIMS SIDS: Natural resources, trading partners, and main industries

	Natural Resources	External Dependence				
		Main Trading Partner		Major Export Commodities	Major Import Commodities	Major Industries
		Import	Export			
Tourism						
Cape Verde	Salt, basalt rock, limestone, kaolin, fish, clay, gypsum	Portugal	Spain	Fuel, shoes, and garments	Food, industrial products, transport equipment	Commerce, transport, tourism, food and beverages, fish processing
Cyprus	Copper, pyrites, asbestos, gypsum, timber, salt, marble	Greece	Greece	Citrus, potatoes, pharmaceuticals, cement, clothing	Consumer goods, petroleum and lubricants, machinery, transport equipment	Tourism, financial services, real estate, food and beverage processing
Maldives	Fish	Singapore	France	Fish	Petroleum products, clothing, intermediate and capital goods	Tourism, fish processing, shipping
Seychelles	Fish, copra, cinnamon tree	Saudi Arabia	France	Canned tuna, frozen fish, cinnamon bark	Machinery and equipment, food products, petroleum products	Fishing, tourism, processing of coconut and vanilla
Agriculture						
Comoros	Negligible	Pakistan	Netherlands	Vanilla, ylang ylang, cloves	Foodstuff, consumer goods, and petroleum	Fishing, tourism, perfume distillation (ylang ylang)
Guinea-Bissau	Fish, timber, phosphates, bauxite, granite, lime stones, deposits of petroleum	Portugal	India	Fish, shrimp, cashew nuts	Foodstuff, machinery, petroleum by-products	Agricultural products processing, beer, soft drinks

Mauritius	Arable land, fish	India	UK	Clothing and textiles, sugar, and cut flowers	Manufactured goods, capital equipment, food stuff	Food processing, textiles, clothing, mining, chemicals
Sao Tome and Principe	Fish, hydropower	Portugal	Netherlands	Cocoa, copra, coffee	Machinery and electrical equipment, food products, petroleum products	Light construction, textiles, soap, beer, fish processing
Hydrocarbon (Petroleum)						
Bahrain	Oil, natural gas, fish, pearls	Saudi Arabia	Saudi Arabia	Petroleum/ by-products, aluminium, textiles	Crude oil, machinery, chemicals	Petroleum processing and refining, aluminium smelting, iron pelletization, fertilizers, off-shore banking
Electronics						
Singapore	Fish, deep water ports	Malaysia	Malaysia	Machinery and Equipment pharmaceuticals, refined petroleum products	Machinery and equipment, mineral fuels, chemicals	Electronics, chemicals, financial services, oil drilling equipment,

Source(s): Data extracted from the Central Intelligence Agency, World Factbook (2013).

small percentage of arable land and fossil fuel reserves, resulting in a high level of economic vulnerability. Three of the least developed countries in the region, namely Comoros, Guinea-Bissau, and Sao Tome are heavily dependent on primary industries such as agriculture to sustain growth and development.

As a group, these SIDS achieved average growth of 2 percent in 2012. However, each subregion and individual country has a unique story of economic success and struggles. This highlights the challenges to regional identification and economic synergy that this regional grouping faces. A persistent period of constricted growth coupled with recurrent fiscal deficits has led to a regional public debt-to-GDP (gross domestic product) ratio of over 70 percent in 2012, with high levels of debt distress being noted in Comoros, the Maldives, and Sao Tome. This has warranted a deeper analysis of public debt accumulation in these small states, with specific attention being paid to the reversal of gains from debt reduction strategies in some countries.

Public Debt of SIDS in AIMS

Over the past decade many SIDS have struggled with increasing levels of public debt. The AIMS SIDS are no different. As small open economies, these countries are disproportionately vulnerable to and impacted by external economic shocks resulting from structural weaknesses inherent to their economies. These factors, along with severe vulnerability to extreme weather conditions, have increased the debt burden of AIMS SIDS. Consequently, the region has experienced rapid debt accumulation and inflated levels of debt servicing which have, in turn, constrained the region's limited fiscal space and curtailed most of the countries' ability to adequately fund development projects. In this context, the recent food–fuel–financial (Triple F) crisis is another phenomenon that contributed to the rapid debt accumulation in the region.

At the end of 2012, 14 SIDS registered public debt-to-GDP ratios in excess of 60 percent (the broadly accepted threshold for sustainable levels of public debt). Among these, six were SIDS in the AIMS region; one of which, Singapore, had debt-to-GDP levels of more than 100 percent (see figure 5.1). The average debt-to-GDP ratio for AIMS SIDS was twice that of the Pacific in 2012, but remained below the debt-to-GDP ratio for the Caribbean small states.

Amid the anemic global economic outlook that ensued in the wake of the Triple F crisis, public debt ratios for AIMS SIDS were on an upward trajectory. Table 5.4 shows the classification of public indebtedness of AIMS SIDS at the end of 2010, a year after the crisis. Of the ten AIMS SIDS, only one, Singapore, was classified as very highly indebted, while four, Cyprus (albeit marginally), Cabo Verde, São Tomé and Principe,

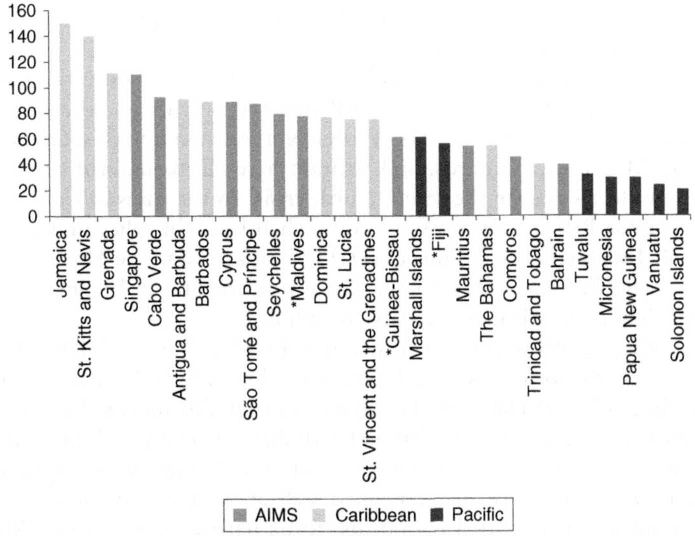

Figure 5.1 Public debt-to-GDP ratio for SIDS as at 2012

Notes: The countries above are UN Member-Small Island Developing States (SIDS), (See www.un.org/specialrep/ohrlls/sid/list.htm). Cyprus is no longer categorised as a Small Island Developing State but is included in the analysis. Countries excluded from Figure 5.1 are Kiribati, Nauru, Palau, Samoa, Tonga and Timor-Leste due to lack of available data.

Source(s): International Monetary Fund 2013e. *Represents IMF estimates.

Table 5.4 Classification of public level of indebtedness, 2010

Moderately (60% or less)	Highly (over 60% to 90%)	Very Highly (over 90%)
Bahrain (36%)	Cyprus (61%)	Singapore (99%)
Comoros (50%)	Cape Verde (75%)	
Mauritius (51%)	São Tomé and Principe (78%)	
Guinea-Bissau (52%)	Seychelles (81%)	
Maldives (60%)		

Notes: The figures in parentheses are the public debt to GDP ratios for each country.

Source(s): International Monetary Fund (2013e).

and Seychelles were classified as highly indebted. The remaining five, Bahrain, Comoros, Guinea-Bissau, Mauritius, and Maldives all had sustainable debt levels of 60 percent or less.

When compared to the 2012 debt levels in figure 5.1, these ratios have increased, some marginally (Guinea-Bissau and Singapore), and others significantly (Maldives and Cabo Verde). Notwithstanding this, Guinea-Bissau, Mauritius, Comoros, and Bahrain have all maintained sustainable debt levels despite the recent global crisis. In the cases of Comoros and Guinea-Bissau, the access to comprehensive debt relief under the Heavily Indebted Poor Countries (HIPC) Initiative and the Multilateral Debt Relief Initiative (MDRI) has been fundamental in keeping their public debt levels within sustainable thresholds.

Another AIMS SIDS which qualifies for HIPC and MDRI is São Tomé and Principe. Surprisingly, despite reaching the HIPC completion point in 2007, São Tomé and Principe remains highly indebted and is at a high risk of external debt distress (IMF 2012b). The IMF's debt sustainability analysis indicated that São Tomé and Principe's present value of debt-to-exports ratio has remained well above the country-specific threshold for an extended period due to its narrow export base (IMF 2012b). Nonetheless, the country has been able to service its current obligations, and given the potential of capitalizing on newly found oil reserves, São Tomé and Principe's medium- to long-term economic outlook may be promising.

Bahrain and Mauritius present interesting case studies for sustainable debt management. Mauritius has legally mandated a public debt ceiling of 60 percent of GDP until 2017 and 50 percent of GDP after 2018 in order to ensure overall debt sustainability.[5] While Bahrain, due to its sound economic fundamentals and booming oil and non-oil sectors, has managed to maintain sustainable public debt ratios during the last decade. More recently, the formation of the Gulf Monetary Union (GMU), which stipulates a public debt-to-GDP ceiling of 60 percent, has helped to further facilitate fiscal stability and low levels of public debt in Bahrain.[6]

Evolution of Debt in the AIMS Region

Prior to the global financial crisis which started in 2008, the public debt profile of selected AIMS SIDS showed: (1) falling public debt from on average 53 percent of GDP in 2002 to 44 percent in 2008; (2) declining external debt service as a percent of exports from on average 5.8 percent in 2003 to 4.9 percent in 2007 for Bahrain, Cabo Verde, Cyprus, Maldives, Mauritius, Seychelles, and Singapore (see figure 5.2).[7] The falling debt ratios over the period were partly driven by international developments, in particular the five-year boom in international commodity

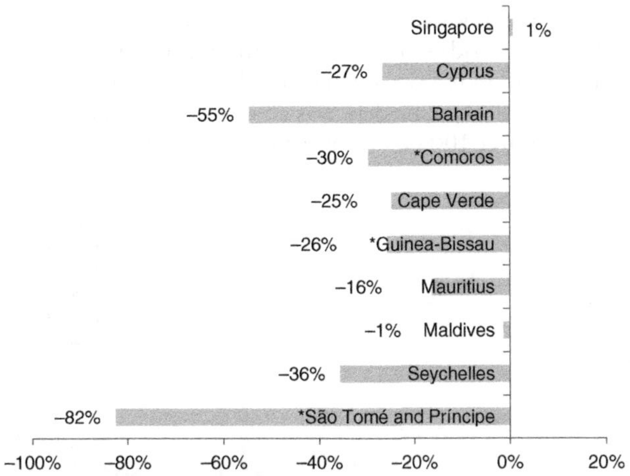

Figure 5.2 Percentage change in debt to GDP ratio of AIMS SIDS between 2002 and 2008

Note: *HIPC and MDRI beneficiary countries.

Source(s): International Monetary Fund (2013e).

prices, which have been favorable to these states, particularly the mineral rich state of Bahrain. For Bahrain, high oil prices have been essential in stimulating economic activity and increasing investments.[8] Additionally, with AIMS SIDS being largely commodity exporters, they were able to benefit from the 138 percent increase in internationally traded food prices in mid-2008.[9] This enabled most of the AIMS SIDS to increase their foreign exchange reserves substantially and reduced the need for additional external financing.

A handful of AIMS SIDS entered the crisis with critically high public debt burdens. The average public debt ratio of Guinea-Bissau, Seychelles and São Tomé and Principe was over 200 percent of GDP up to 2006.[10] The accumulation of public debt in these economies is attributed to various factors. For the Seychelles, significant debt accumulation occurred between 1997 and 2002. This was as a result of excessive spending, particularly in the social sector, and significant borrowing to finance growth during that period.[11] In addition, Seychelles has a historical trend of unsustainable public debt and difficulty repaying its loans due to structural problems. Consequently, the public debt-to-GDP ratio remained well above 100 percent between 2002 and 2007. These occurrences eventually led to the government of Seychelles defaulting on its debt obligations in 2004.

Since then, the government has pursued several strategies to set the economy on a sustainable growth path. The most notable were the macroeconomic reform measures implemented under the guidance of the

IMF.[12] These were followed in 2007 by the launching of the "Seychelles Strategy 2017," a broad-reform initiative to reduce government debt to 60 percent of GDP in ten years while doubling per-capita income. As of June 2011, Seychelles' public debt was an estimated 70 percent of GDP, compared to the over 100 percent it recorded in 2008.

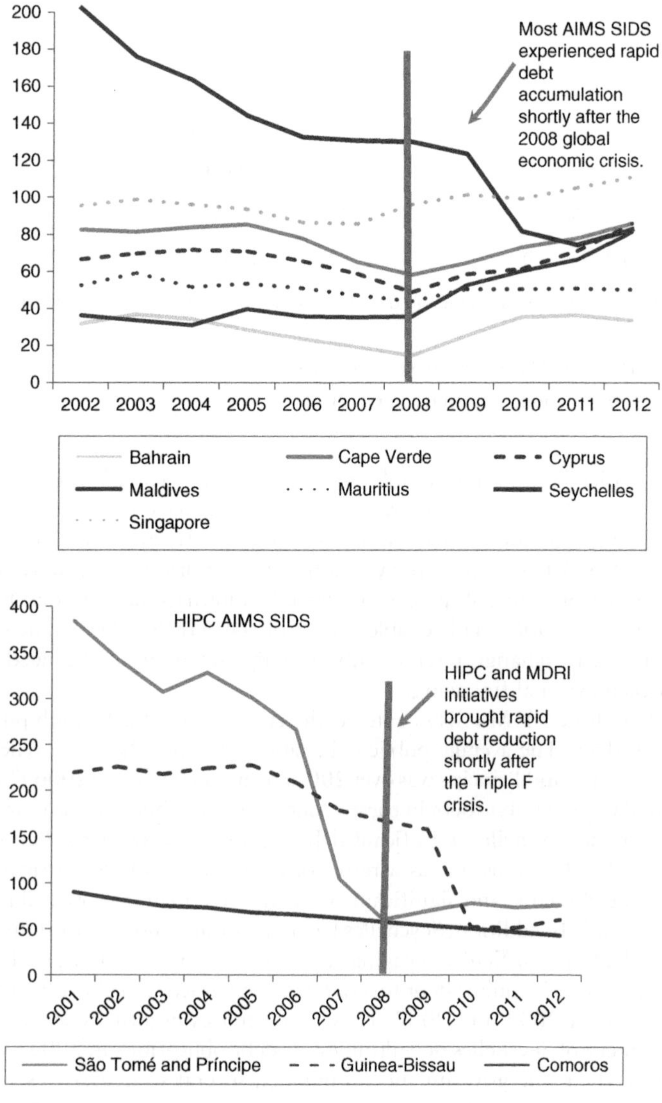

Figure 5.3 Public to GDP ratios of selected AIMS SIDS, 2002–2012

Source(s): International Monetary Fund (2013e).

For Guinea-Bissau, despite having qualified for comprehensive debt relief under the HIPC Initiative, the country did not benefit from most of the debt relief that was committed at its 2000 decision point, as their corresponding Poverty Reduction and Growth Facility (PRGF) program was derailed shortly after being implemented.[13] As a result, Guinea-Bissau could not service its debt in full beyond 2000 and accumulated arrears on most of its debt owed between 2000 and 2006. Comoros on the other hand, received debt relief under HIPC in 2009, but experienced prolonged periods of political instability which brought about gradual deterioration of its institutions, infrastructure, and economy, increasing the need for borrowing.

The Triple F crisis saw a worsening of economic conditions in all these countries, and those which qualified for debt relief (Comoros, Guinea-Bissau, and São Tomé and Principe) received substantial debt cancellation under the HIPC and MDR Initiatives. As a result of those initiatives, these countries were left with lower debt burdens after the global economic crisis. As clearly illustrated in figure 5.3, this represents a marked difference in the debt trajectory compared to those AIMS SIDS that did not benefit from the debt relief initiatives.

Overall, prior to the global financial crisis, the debt burdens of AIMS SIDS exhibited mostly downward trends. The global economic crisis however worsened economic conditions in most of the countries, with a period of rapid debt accumulation ensuing shortly after 2008 in six of the ten AIMS SIDS. Three of the four SIDS that did not experience this rapid post-crisis accumulation of debt benefitted from HIPC debt relief efforts. The fourth (Seychelles) benefitted from IMF-supported macroeconomic reform programs. Rapid debt accumulation in the non-HIPC AIMS SIDS in a crisis context of reduced Overseas Development Assistance (ODA) and concessional financing, suggests that these countries may have to become more reliant on market-based non-concessional financing to support economic development and meet fiscal needs.[14]

Debt Service

Of the countries for which data are available, AIMS SIDS have generally maintained low debt service ratios over the last decade.[15] Between 2000 and 2005, total debt service averaged 11 percent of exports, just below the sustainable threshold of 15 percent. Maldives registered the lowest debt service ratio of roughly 5 percent over the period. By contrast, São Tomé and Principe was the only AIMS SIDS to breach the 15 percent benchmark between 2000 and 2005, with total debt service ratio amounting to 39 percent in 2004. This, along with various other structural deficiencies ensured that the country qualified for debt relief under HIPC. But despite completing the HIPC Initiative in 2007, the debt service ratio remained extremely high at 31 percent. Only after receiving additional debt relief from the MDRI did total debt service decline to sustainable levels (as low as 5 percent of exports in 2011).

The latter half of the decade generally saw improved external debt service ratios for the AIMS SIDS, with average external debt service of 8 percent of exports between 2006 and 2011. The outliers during this period were Comoros, with external debt servicing obligations at 37 percent of exports in 2007 and Seychelles, with interest payments on public debt amounting to 32 percent of government revenues in 2009. Given the deteriorating situation in Seychelles, the government had to restructure its bilateral and private external debt to help alleviate the external debt burden.[16] This restructuring, along with the reform measures currently being implemented, have significantly improved the sustainability of the country's debt.

Composition of Debt: Domestic versus External

Domestic debt levels are relatively high among AIMS SIDS. As seen in figure 5.4, the majority of the countries have higher domestic debt relative to external debt. Singapore stands out among AIMS SIDS with a debt stock comprising solely of domestic creditors and has maintained this trend for the last five years. For the Maldives, however, increases in domestic debt are a recent phenomenon. Between 2004 and 2010, domestic debt increased rapidly from just 15 percent of GDP in 2004 to 54.5 percent in 2010. As at 2012, domestic debt in the Maldives stands at 60 percent of GDP. According to the IMF, the recent increases in domestic debt may be attributed to the implementation of countercyclical fiscal

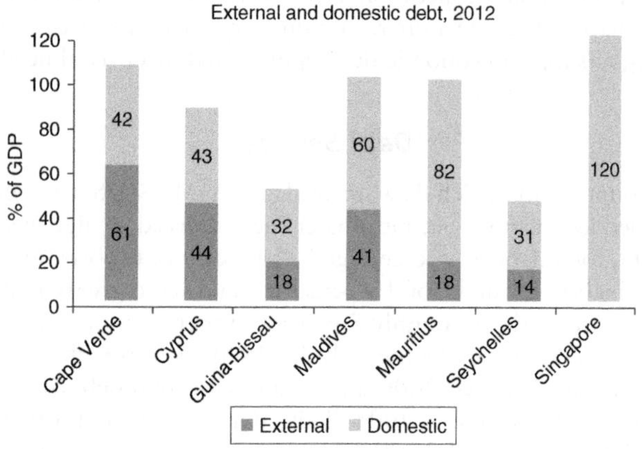

Figure 5.4 Composition of public debt for selected AIMS SIDS

Source(s): International Monetary Fund Article IV, various reports.

Table 5.5 Standard and Poor's foreign currency rating for AIMS SIDS

AIMS SIDS	Rating
Bahrain	BBB
Cape Verde	B
Cyprus	B
Singapore	AAA
Other AIMS SIDS	Not reported

Source(s): Standard and Poor's Rating Services (2014).

policies in response to the global economic crisis. In fact, they reported that for low-income countries, the financial and economic crisis increased domestic debt ratios by on average 3 percent of GDP in 2009 and 2010. Among the AIMS SIDS, Cabo Verde, Guinea-Bissau, and Seychelles had increases in domestic debt levels of around 3 percent after the crisis.

For the remaining AIMS SIDS, the build-up in domestic debt was due to their well-developed bond markets. Countries such as Mauritius, Bahrain, and Singapore made significant use of domestic bond issuance to offset deficits which led to a significant portion of their debt being domestic debt rather than external.[17] Furthermore, having a relatively stable investment grade status has helped to bolster investor confidence in these economies, thus reducing their need for external borrowing (see table 5.5).

The gradual shift in composition of the debt, from external to domestic has serious implications in terms of the vulnerability it poses to debt in these sovereign states. While increases in domestic debt may have the advantage of carrying little exchange risk (if bonds are denominated in domestic currency), domestic debt instruments are usually more costly relative to external finance given the shorter maturity and higher interest rates of domestic instruments. For instance, in Seychelles, the interest rate on domestically issued sovereign bonds in 2009 was as high as 12.97 percent, the highest among all AIMS SIDS.[18] Domestic debts are also usually more difficult to restructure when there are repayment difficulties, and pose a significant threat to domestic banks and financial institutions in the event of a sovereign default. This implies that shifts in the composition of debt toward domestic sources can increase debt vulnerabilities for governments.

Current Factors Affecting Debt Trends in the AIMS SIDS

Concessional Financing and Debt Restructuring in AIMS SIDS

Concessional financing has been an alternate source of funds for those AIMS SIDS that qualified for this type of financing. However, concessional financing has decreased over the past decades and will continue to decrease given the increased focus on Least Developed Countries.[19] ODA flows to AIMS SIDS have declined significantly over the periods 1960–2000; 2001–2005 and 2007–2011 (see table 5.6). For Cabo Verde, a low-income, highly indebted country, ODA has decreased by 83 percent between 2001–2005 and 2010. Similarly, the highly indebted countries of Comoros, Guinea-Bissau, and São Tomé and Principe experienced reductions in ODA of 84, 88, and 89 percent, respectively over the same time period. Notwithstanding these reductions, concessional financing has proved to be effective in reducing the debt levels in the four low-income AIMS SIDS; namely Cabo Verde, Comoros, Guinea-Bissau, and São Tomé and Principe. This suggests concessional finance may be an appropriate mechanism to help AIMS SIDS deal with their growing debt burdens.

Debt restructurings have also been essential in reducing the overall debt levels in the AIMS region. Table 5.7 provides a sample of the debt restructuring undertaken across the region since 1980. With the exception of Bahrain, Mauritius, and Singapore, the remaining seven AIMS SIDS have concluded at least 24 restructuring operations over the last three decades, involving multilateral, private external (Paris Club), and domestic creditors. Between 2009 and 2013, Comoros, Cyprus, Guinea-Bissau, and Seychelles all entered in debt restructuring agreements.[20]

Although Cabo Verde's recent graduation from LDC status precludes debt relief from traditional multilateral sources, a historical analysis of the country's performance under debt restructurings indicates that such initiatives were essential to achieving sustainability. Bilateral debt relief

Table 5.6 AIMS SIDS: Official development assistance, US$ '000

	1960–2000	2001–2005	2006	2007	2008	2009	2010	2011	Total Aid
Bahrain	2959	322	0	0	0	0	0	0	3282
Cape Verde	2089	1921	138	165	222	196	328	251	5310
Comoros	1078	417	32	45	42	50	67	52	1782
Cyprus	986	0	0	0	0	0	0	0	986
Guinea-Bissau	2224	1161	87	122	134	147	139	119	4133
Maldives	521	499	38	37	54	33	111	46	1339
Mauritius	1153	759	19	69	110	155	125	183	2573
Sao Tome and Principe	692	444	23	51	47	31	49	75	1412
Seychelles	537	245	14	11	13	23	56	21	919
Singapore	603	0	0	0	0	0	0	0	603
Total AIMS Aid	12842	5768	351	500	622	635	875	747	22339

Source(s): World Bank (2013b).

Table 5.7 The frequency of debt restructurings in the AIMS region, 1980–2013*

Country	Multilateral Debt	Year	Type	Paris Club Debt	Year	Type	Commercial Debt	Year	Type	Public Domestic Debt	Year	Type
Bahrain	–	–	–	–	–	–	–	–	–	–	–	–
Cape Verde	UK MDRI	2006	Other	–	–	–				✓	1998	Debt Exchange
Comoros	Full debt relief under HIPC & MDRI	2012	Cancellation	✓	2009 2010 2013	Cancellation & Rescheduling	✓	2010	Rescheduling	–	–	–
Cyprus										✓	2013	Debt Exchange
Guinea-Bissau	Full debt relief under HIPC & MDRI**	2010	Cancellation	Paris Club (Ad hoc)	1987 1989 1995 2001 2010 2011 2013	Cancellation & Rescheduling	✓	1995	Debt Swap	–	–	–
Maldives	–	–	–	–	–	–	–	–	–	✓	2006	Debt Exchange
Mauritius	–	–	–	–	–	–	–	–	–	–	–	–
Sao Tome & Principle	Full debt relief under HIPC & MDRI	2008	Cancellation	✓	2000 2005 2007	Cancellation & Rescheduling	✓	1994	Debt Buyback	–	–	–
Seychelles	–	–	–	Paris Club (Ad hoc)	2005 2009	Cancellation, Moratorium & Rescheduling	✓	2009	Debt Exchange	–	–	–
Singapore	–	–	–	–	–	–	–	–	–	–	–	–

Notes: *Data Sheet does not include IMF Loans. Brazil has announced (in 2013) a plan to restructure the debt for select African countries. Yet to materialize.
**In 2009 the Maldivian government moved away from the monetization of its debt to the introduction of sovereign bonds. The first set of bonds issued in 2009 totalled $640 million Rufiyaa. Does not include topping up and Guinea Bissau had additional debt cancelled bilaterally, including loans from Cuba.

Source(s): Various articles from Club de Paris, World Bank and the IMF for various years, Maldives Monetary Authority (2013), OECD (2011).

contributed to a steady reduction in the country's debt service ratio, falling from 29.2 percent in 2001 to 10.6 percent in 2003.

The debt restructuring in Seychelles similarly proved to be effective in reducing the country's high debt levels. Against the background of weak economic performance, upward spiraling inflation, and balance of payments problems, Seychelles negotiated a comprehensive restructuring of its official bilateral debt on Evian terms with its Paris Club creditors in April 2009 and subsequently launched a debt exchange offer to further ensure debt sustainability.[21] The result of these initiatives was a 50 percent nominal reduction in its external commercial debt, reduction in debt to GDP from 123 percent at end 2009 to 81 percent at end 2010 (the lowest ratio recorded in Seychelles over the last two decades), and a significant improvement in credit worthiness.[22]

Seychelles' outcome presents a tempting case for utilizing debt restructurings/exchanges in the absence of comprehensive debt relief to help alleviate unsustainable debt burdens.

The 2008–2009 Global Food, Fuel, and Financial Crisis: Impact and Recovery

The years preceding the Triple F Crisis saw regional economic growth reaching 5 percent of GDP in the 1990s, peaking at 7.8 percent in 2006 (see table 5.8). Growth during the crisis slipped by 5.7 percent between 2007 and 2009. Seychelles and Maldives were the hardest hit by the global recession, with growth rates of the former falling from 10 percent in 2007 to –2.1 percent in 2008, and the latter falling from 12 percent in 2008 to –3.6 percent in 2009.[23]

The impact of the recession on key revenue sources such as tourism, remittances, and trade led to a sharp increase in the current account deficit of most the countries. Both Bahrain and Singapore maintained net exporting statuses although the Triple F crisis caused an 81 percent and 28 percent decrease in their respective current account surpluses. The resulting strain on revenues and the increased expenditure to stimulate demand in the region reversed the generally decreasing public debt trajectory, as the onset of the crisis saw the return of increasing debt levels to Bahrain, Cabo Verde, Cyprus, Maldives, and Sao Tome and Principe.[24]

The region rebounded from the Triple F crisis with a sharp return to growth, underpinned by the recovery of tourism and remittance inflows. Individual countries, however, remain at different stages of recovery as Bahrain and São Tomé and Principe achieved growth above 4 percent between 2011 and 2012, while growth rates were reduced by 3 percent in Mauritius. Singapore's robust growth of 14.8 percent immediately following the recessionary period also tapered off to a

Table 5.8 AIMS SIDS: Real GDP growth, 1981–2012

	1981–1990	1991–2000	2001–2005	2006–2010	2006	2007	2008	2009	2010	2011	2012
Atlantic	2.1	3.3	3.7	4.8	8.0	4.8	6.3	1.9	3.2	4.8	1.7
Cape Verde	5.3	6.8	5.9	5.0	9.1	9.2	6.7	-1.3	1.5	4.0	2.5
São Tomé and Principe	-1.4	1.5	3.6	6.4	12.6	2.0	9.1	4.0	4.5	5.0	4.0
Guinea-Bissau	2.4	1.5	1.6	3.0	2.1	3.2	3.2	3.0	3.5	5.3	-1.5
Indian Ocean	**4.6**	**4.8**	**3.5**	**5.1**	**8.3**	**7.1**	**4.6**	**0.9**	**4.7**	**3.9**	**3.0**
Bahrain	3.6	4.7	6.1	5.9	6.7	8.4	6.3	3.2	4.7	2.1	4.8
Maldives	8.2	7.5	5.5	9.2	19.6	10.6	12.2	-3.6	7.1	6.5	0.9
Mauritius	4.9	6.1	3.3	4.6	4.5	5.9	5.5	3.0	4.1	3.8	3.3
Seychelles	3.7	4.6	-0.2	4.6	9.4	10.1	-1.9	-0.2	5.6	5.0	2.9
Comoros	2.7	1.2	2.8	1.3	1.2	0.5	1.0	1.8	2.1	2.2	3.0
Mediterranean											
Cyprus	6.3	4.6	3.2	2.5	4.1	5.1	3.6	-1.9	1.3	0.5	-2.4
South China Sea											
Singapore	7.8	7.2	4.8	6.7	8.6	9.0	1.8	-0.8	14.8	5.2	1.3
AIMS Average	**4.4**	**4.6**	**3.7**	**4.9**	**7.8**	**6.4**	**4.8**	**0.7**	**4.9**	**4.0**	**1.9**

Source(s): International Monetary Fund (2013e).

mere 1.3 percent in 2012. Having experienced positive growth after the Triple F crisis, Guinea Bissau and Cyprus also experienced recessionary conditions in 2012. However, improvements in Guinea Bissau's political environment and access to concessional loans are expected to boost growth by 3.5 percent while Cyprus is expected to remain in a recession (International Monetary Fund 2013a, 2013b).

The current concerns of slowing regional growth and increases in public debt have warranted a deeper analysis of key revenue earners that were affected by the recession. The post crisis impact of these factors on the region's growth and public debt will also be assessed as the ability to generate revenue is at the crux of debt reduction strategies.

Trade

In 2009, regional exports declined by 8.3 percent. There were noticeable declines in all countries, except for Comoros and Seychelles which benefitted from IMF driven exchange rate devaluations (see figure 5.5).[25] The consequent decreases in foreign exchange inflows reduced the debt servicing capacity of the region during the crisis. Overall, the region experienced a 5 percent increase in debt between 2008 and 2009 as reduced international demand deflated the region's income and subsequent government revenues. This contributed to worsening fiscal imbalances that increased the need for debt financing for the AIMS SIDS.

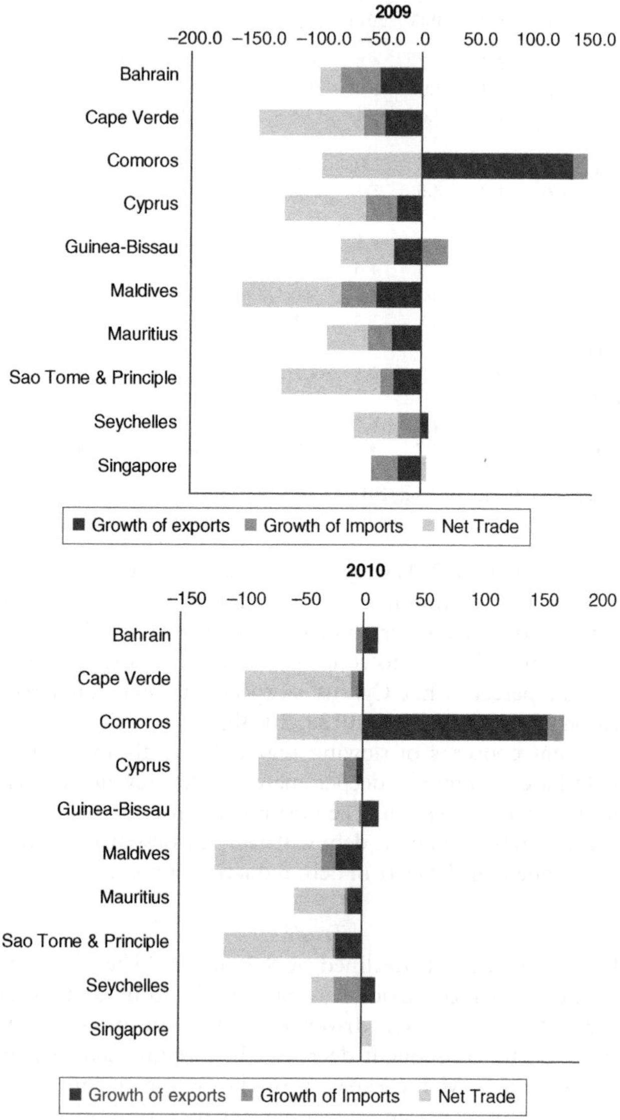

Figure 5.5 AIMS SIDS: Terms of trade

Source(s): International Trade Centre (2013).

Tourism and Remittances

More than half of AIMS SIDS are highly tourism based economies (HTBEs), making them vulnerable to fluctuations in tourism-related earnings.[26] Tourism revenues, one of the chief sources of foreign exchange for debt servicing, received a blow in 2009 with a US$3 billion reduction from the previous year. This was largely due to a regional decline in tourist arrivals from 23.2 million to 22.8 million people between 2008 and 2009 (see table 5.9). The largest decline was seen in Singapore and Cyprus, whereas Bahrain and Cabo Verde continued to record strong numbers throughout the recession and into 2010. In the case of Bahrain the resilience is attributed to the fact that the majority of the country's visitors were from Saudi Arabia—a country that was not affected by the recessionary pressures.

The postcrisis recovery of the tourism industry remains positive, as the sector achieved revenues of US$22 billion in 2010, with most AIMS SIDS—except for Cyprus, Comoros, and Bahrain—exceeding their 2008 figures. As the region struggles with increasing debt levels the rebound of tourism has made major contributions to regional growth and revenues. It is expected that the recovery of the tourism sector will reduce the need for additional financing.

The AIMS region is home to three of the world's top ten remittance receiving small states—Mauritius, Cyprus, and Cabo Verde. The average remittance to GDP ratio for the AIMS region is 3.9 percent, which is five times the world average.[27] Furthermore, remittance as a percent of trade in goods and services for the AIMS region is 5 percent higher than the average for SIDS. Europe contributes over 79 percent of the region's remittances followed by the United States (US) with 6.4 percent.[28] These

Table 5.9 AIMS SIDS: Tourist arrivals, 2006–2011

AIMS SIDS	Pre-Crisis		Crisis Period		Post-Crisis	
Country	2006	2007	2008	2009	2010	2011
Bahrain	7,289,000	7,833,000	8,631,000	8,861,000	11,952,000	6,732,000
Cape Verde	242,000	267,000	285,000	287,000	336,000	428,000
Comoros	29,000	15,000	15,000	11,000	–	–
Cyprus	2,401,000	2,416,000	2,404,000	2,141,000	2,173,000	2,392,000
Guinea-Bissau	12,000	30,000	–	–	–	–
Maldives	602,000	676,000	683,000	656,000	792,000	931,000
Mauritius	788,000	907,000	930,000	871,000	935,000	965,000
Sao Tome	12,000	12,000	14,000	15,000	8,000	–
Seychelles	141,000	161,000	159,000	158,000	175,000	194,000
Singapore	7,588,000	7,957,000	7,778,000	7,489,000	9,161,000	10,390,000
AIMS SIDS	**21,014,400**	**22,301,400**	**23,221,111**	**22,765,556**	**28,723,500**	**25,179,429**

Source(s): World Bank (2013b).

Table 5.10 Percent changes in personal remittances, 2006–2011

Selected AIMS SIDS	2006–2007	2007–2008	2008–2009	2009–2010	2010–2011	2011–2012
Cape Verde	1.66	11.69	-11.84	-4.24	34.99	-5.68
Cyprus	36.08	64.81	-45.80	6.57	-6.30	-11.35
Guinea-Bissau	68.62	14.94	-1.23	-6.06	n.a	n.a
Maldives	183.46	-21.03	-28.01	-29.96	-5.12	5.00
Mauritius	-3.46	-9.93	4.22	23.30	-36.27	66.08
Sao Tome and Principe	25.00	50.00	-33.33	218.16	8.14	-7.59
Seychelles	-94.76	-52.16	286.14	12.46	42.17	-37.18
Change in Total Remittances	**17.68**	**33.90**	**-28.52**	**0.91**	**-2.45**	**-7.98**

Source(s): World Bank (2013a).

small states are thus heavily dependent on economies that were hardest hit by the Triple F crisis.

The record levels of unemployment and financial challenges in Europe and the US contributed to a 28.5 percent decrease in remittances for AIMS SIDS between 2008 and 2009 (see table 5.10). The countries that experienced the largest decreases in remittances were the Maldives, São Tomé and Principe, and Cyprus. Although the region experienced positive growth in remittances immediately after the recession, the uncertainties in the global economy, especially in Europe, further challenged the region's migrants resulting in a return to declining remittances between 2011 and 2012.

The World Bank's Migration and Remittance publication revealed that remittances had a significant effect on household income, indirectly and directly relieving the need for social security transfers (Adams 2006). In addition, the tourism dependence of the region makes this service a vital part of the composition of revenues. Therefore, the temporary reductions in tourism and remittances during the 2008–2009 Crisis resulted both in the loss of tax revenues and increases in government expenditure to subsidize the loss of demand due to the reduced income of households. This contributed to the increase in fiscal deficits and public debt between 2007 and 2009. Consequently, the recovery of these sectors after the recession was crucial to the return of growth to the region in 2010.

Political Instability: The Experiences of Guinea-Bissau, Comoros, and Bahrain

An uncertain political climate remains a major challenge to debt sustainability in some countries in the AIMS region, namely Guinea-Bissau,

Comoros, and Bahrain. The most politically unstable and poorest country in the region is Guinea-Bissau.[29] Over the years, political conflicts in this country have been violent, emanating from 13 years of guerrilla warfare prior to independence from Portugal in 1974. The violence continued into the 2000s with internal unrest and military coups. This political insecurity threatens debt sustainability and investor confidence following the HIPC and MDR Initiatives.

Of note is the 2012 military coup which crippled public institutions and forced the closure of harbors and airports (World Bank 2014). This adversely affected the importation of crucial agricultural inputs such as seeds and fertilizers and reduced cashew exports, the country's main earner. This contributed to a contraction of GDP by 1.5 percent in 2012 which is in contrast to growth of 3.5 and 5.3 percent in 2010 and 2011, respectively (International Monetary Fund 2014c). Moreover, this internal conflict forced overseas donors to reduce aid to the politically unstable nation. This withdrawal of aid has constrained the country to accessing less favorable loans from the African Development Bank which has increased the share of regional financing from 0.1 percent to 21 percent of total public debt (International Monetary Fund 2013b).

In Bahrain and Comoros, tourism is a major revenue earner that is vulnerable to the unstable political environment in these countries. Bahrain's political conflicts in 2011 resulted from a call for political reform. This upsurge in violence resulted in a 77 percent decrease in tourist arrivals (recall table 5.9) and the cancellation of the Formula One Grand Prix which caused potential losses of US$600 million. Bahrain's non-oil growth is also projected to remain below 4 percent over the medium term due to continued political instability in the wake of calls for government reform. Over 20 coups have been noted in the Comorian islands since 1975, with 2 of the islands declaring independence through violent means as recently as 1997. Competition from other tourist destinations such as Mauritius and the Seychelles continue to threaten the Comorian tourism sector as waves of political instability make the country a less favorable destination.

The direct impact of political instability on the public debt of Guinea- Bissau and Comoros can be seen in table 5.11. Both countries have requested emergency assistance from the IMF totaling US$14.9 million to fund post-conflict rehabilitation; with Guinea-Bissau exceeding half of its quota in 2008. The impact of political instability on the debt of Guinea-Bissau, Comoros, and Bahrain should act as a lesson to the region, as the growing political uncertainty and disagreements in other countries such as the Maldives, Sao Tome, and Cyprus can escalate this problem into a regional threat to debt sustainability.

Table 5.11 AIMS SIDS: Post-conflict emergency assistance from the IMF

	Year	Event	Amount (US million)	Percent of Quota
Guinea-Bissau	1999	Civil war	2.9	15.0
	2000	Civil war	1.9	10.0
	2008	Military coup	5.7	25.0
	2008	Military coup	2.7	12.5
Comoros	2008	Military coup	1.7	12.5

Source(s): Various IMF articles for various years.

The Impact of Natural Disasters

Natural disasters such as floods, droughts, tsunamis, and volcanic activity as well as recent epidemic threats of cholera and influenza have affected the economic livelihood and infrastructure of the AIMS SIDS.[30] Specific countries in the region such as Seychelles, Maldives, Cabo Verde, and Comoros have experienced recent natural disasters. These include flooding in Comoros and the Seychelles, with the latter enduring numerous floods during the 2000s. Heavy rains in 2012 also devastated the Comorian islands costing the country over US$20 million in physical damages (GFDRR 2013).[31]

One of the largest natural disasters that had a noticeable impact on debt levels was the 2004 Asian Tsunami. The tsunami impacted both the Seychelles and the Maldives, with the devastation being greatest in the Maldives. Damage to the Maldive's infrastructure resulted in losses of over 62 percent of GDP, and a 5.5 percent reduction in income (UNDP 2010b). Public debt also increased rapidly in light of post-tsunami reconstruction efforts, escalating from 31 percent of GDP in 2004 to 81 percent in 2012 following emergency assistance from the IMF.[32] The Seychelles' graduation to middle-income status in 1997 meant that it no longer qualified for systematic program support and concessionary loans. This increased the expenditure required during the recovery process and contributed to a 13 percent increase in debt between 2003 and 2004.

Table 5.12 gives a representation of some of the major natural disasters that have impacted the region. There is no regional facility that collates data on damage estimation and the sources used to fund recovery, which would serve to inform future disaster mitigation efforts. However, a look at the movement in aid and debt during the concurring periods demonstrates that in many instances there was an increase in debt for the period directly following the disaster as well as an increase in aid to offset expenditures. With the increasing frequency and intensity with which these disasters have been occurring, the socioeconomic consequences for the AIMS SIDS may be long lasting. In fact, the United Nations Conference on Trade and Development (UNCTAD) estimated

Table 5.12 AIMS SIDS: Selected natural disasters, 2000–2013

Country	Year	Disaster	Amount of People Affected	Percent Change in Aid and Debt in year following the disaster	
				Aid	Public Debt
Bahrain	–	–	–		
Cape Verde	2002	Drought	30000	60.9	7.8
	2009	Flood	Not reported	67.6	15.2
	2009	Epidemic	20147		
Comoros	2005	Volcanic eruption	245000	39.7	0.4
	2007	Epidemic	1490	41.1	-0.9
	2009	Flood	2500	-7.4	0.1
	2012	Flood	65137		–
Cyprus	2000	Extreme temperature	400		10.7
	2003	Storm	30		10.8
	2004	Storm	10		5.9
	2007	Extreme temperature	Not reported		-1.5
Guinea-Bissau	2004	Flood	Not Reported	-13.6	0.5
	2004	Epidemic	Not Reported		
	2005	Epidemic	Not Reported	31.8	-1.0
	2006	Drought	32000	40.6	-4.3
	2008	Epidemic	14004	9.5	8.6
	2010	Flood	56792	-14.8	-2.1
Maldives	2004	Earthquake	27214	189.1	18.0
	2007	Flood	1649		24.5
	2011	Epidemic	1289		25.8
Mauritius	2002	Storm	1050	-159.9	25.9
	2006	Epidemic	2553		6.0
	2007	Storm	Not Reported	59.2	4.6
	2013	Flood	Not Reported	–	
Sao Tome	–	–	–		
Seychelles	2002	Storm	6800	-82.4	-12.3
	2004	Earthquake	4830	2.1	-3.3
	2006	Epidemic	5461	-18.5	21.8
	2013	Storm	3000	–	–
Singapore	2000	Epidemic	2022	–	10.9
	2003	Epidemic	205	–	10.2

Source(s): Centre for Research on Epidemiology of Disasters (CRED) (2013) and International Monetary Fund (2014c).

that natural disasters which struck low-income SIDS between 1980 and 2008, added on average 24 percentage points to the debt-to-GDP ratio (UNDP 2010a). This results from the inability of small states to marshal the substantial resources needed in the aftermath of a major disaster.

Fiscal Deficits

Sustained fiscal deficits have an adverse effect on a country's debt levels. Such deficits have been identified as impacting the current debt ratios of many of the AIMS SIDS. The fiscal positions of selected countries within their respective subregions are assessed below.

The Atlantic

The SIDS in the Atlantic have maintained, on average, consistent fiscal deficits since 2008. For some countries this has been attributed mainly to the expansion of public investment initiatives, in an attempt to induce long-term growth. Table 5.13 highlights significant increases in capital expenditure between 2008 and 2011. In other countries the reduction of concessional funding contributed to the fiscal deficits, as indicated by the dramatically declining grants to GDP ratio between 2008 and 2012. In some countries, such as Cabo Verde, Sao Tome and Principe, and Guinea-Bissau, a coincidence of both these factors precipitated widening fiscal deficits and debt accumulation.

Cabo Verde, for example, received a five-year window for the discontinuation of concessional funding subsequent to graduation from LDC status in 2007. The government therefore seized the opportunity to use concessional loans to engage in a large scale public investment program to support domestic demand, and raise GDP growth. An increase in capital expenditure of 39 percent between 2009 and 2010 was accommodated by a 13 percent increase in public debt. This initiative has been successful as growth returned to the economy peaking at 4 percent in 2011. It was however costly. Public debt to GDP rose from 65 percent in 2007 to 86 percent in 2012. It is expected that the increasing trajectory of public debt will be partially reduced by continued reductions in domestic debt. The fiscal prudence of the Cabo Verdean government has also urged development partners to consider extending the transitional window to further support the scaling up of public investment (International Monetary Fund 2012a). This, however, will be negatively impacted by the continued accrual of contingent liabilities. In Cabo Verde, contingent liabilities associated with debt reached 11 percent of GDP in 2010.

São Tomé and Principe's fiscal balance has been heavily sustained by grants, with the HIPC debt relief and the subsequent benefits from the HIPC/MDRI resulting in an abnormal fiscal surplus of 125 percent of GDP in 2007. Such grants have, however, subsequently declined. With the anticipation of oil production in the near future, capital expenditure has increased to facilitate the introduction of the requisite amenities to improve the local business environment, including the building of a deep-water port scheduled for commencement in 2014. As a result of this, total public debt to GDP has increased, reaching 77 percent in 2012 (International Monetary Fund 2012b).

Table 5.13 Atlantic SIDS: Central government operations, 2001–2012

	2001–2005	2006	2007	2008	2009	2010	2011	2012
	Average				Total			
(US$, millions)								
Revenue and Grants	352.0	514.6	830.2	759.0	731.9	726.2	802.1	699.6
Tax revenue	190.0	318.3	390.5	467.0	409.0	420.3	503.1	448.8
Grants	106.6	114.2	119.5	124.8	119.7	119.5	119.5	120.6
Other Revenue	55.4	82.1	320.2	167.2	203.2	186.4	179.5	130.2
Expenditure	404.9	582.5	648.5	790.8	830.5	916.8	1033.6	916.9
Capital expenditure	146.0	216.2	241.6	299.6	357.9	443.7	489.6	443.8
Current expenditure	258.9	366.3	406.9	491.2	472.6	473.1	544.0	473.1
Wages and salaries	114.8	155.4	174.8	210.0	245.3	266.4	298.4	250.6
Fiscal Balance	–52.9	–68.0	181.7	–32.0	–98.6	–190.6	–231.5	–217.3
(Percent of GDP)								
Revenue and Grants	29.4	27.9	70.8	30.6	27.8	29.2	28.0	24.6
Grants	10.8	9.1	44.4	15.6	13.7	8.4	9.1	9.3
Expenditure	33.9	34.0	29.4	27.1	34.1	36.5	35.7	32.0
Fiscal Balance	–4.2	–6.0	41.4	3.5	–6.3	–7.3	–7.7	–7.4

Source(s): African Development Bank (2014).

In Guinea-Bissau, a five-year period of steadily increasing grant inflows was followed by a sharp decrease in grants (32 percent) between 2011 and 2012. Government expenditures simultaneously increased, with the need to finance food security initiatives, infrastructural investments in roads, and rice production (International Monetary Fund 2013b). The consequent increase in the country's fiscal deficits was met with attempts to contract capital expenditures, which decreased by 10 percent between 2001 and 2012.

Indian Ocean

The similarities between the SIDS in the Indian Ocean are less apparent than for those in the Atlantic, as this grouping of islands has had diverse socioeconomic and fiscal experiences. Bahrain and the Maldives have had the highest rate of public debt growth in the Indian Ocean since 2009. As such, deeper scrutiny into the cause of debt accumulation in these economies will ensue.

After a small deficit in 2002, Bahrain accumulated surpluses of approximately US$5 billion between 2003 and 2008. Public debt to GDP also decreased by 20 percent during the period mentioned. The impact of the decline in oil prices due to the recession became apparent in 2009 with the deterioration of the country's fiscal situation. The fiscal surplus of 4.9 percent of GDP eroded, resulting in a fiscal deficit of 6.6 percent of GDP (see table 5.14). Consequently, the debt-to-GDP ratio rose from 14.6 percent in 2008 to 35.5 percent in 2010. The increase in the deficit

Table 5.14 Indian Ocean SIDS: Fiscal balance

		2006	2007	2008	2009	2010	2011	2012
		(Percent of GDP)						
Comoros	Revenue	18.7	20.3	23.5	23.6	29.2	23.6	28.9
	Expenditure	21.3	22.3	26.0	23.0	22.2	22.1	25.5
	Fiscal Balance	−2.6	−2.0	−2.5	0.6	7.0	1.4	3.4
Seychelles	Revenue	41.0	32.0	34.9	37.0	35.0	38.2	42.6
	Expenditure	47.1	41.5	29.4	34.1	35.8	35.7	40.2
	Fiscal Balance	−6.1	-9.5	5.5	2.8	−0.8	2.5	2.4
Mauritius	Revenue	19.1	19.6	21.0	22.8	21.9	21.5	21.4
	Expenditure	23.5	22.8	23.8	26.3	25.1	24.7	23.3
	Fiscal Balance	−4.4	-3.3	−2.8	-3.6	−3.2	−3.2	−1.8
Bahrain	Revenue	30.9	29.3	32.4	23.8	27.2	29.2	29.9
	Expenditure	28.2	27.5	27.5	30.5	34.2	30.9	32.5
	Fiscal balance	2.7	1.9	4.9	−6.6	−7.0	−1.7	−2.6
Maldives	Revenue	36.6	38.2	30.6	22.5	23.8	30.8	30.3
	Expenditure	42.1	42.0	42.5	43.7	40.1	43.1	43.8
	Fiscal Balance	−5.5	−3.8	-11.9	-21.1	-16.3	-12.3	-13.5

Source(s): International Monetary Fund (2014c).

emanated from the shortfall in revenues, which are heavily dependent on oil and gas exports, between 2008 and 2009. Revenues from hydrocarbons are important to fiscal stability being at least 80 percent of total government revenues since 2007 (Bahrain Economic and Development Board 2013).

Government expenditure grew by 3 percent of GDP between 2008 and 2009 and 5 percent between 2008 and 2012 (Bahrain Economic and Development Board 2013). Capital expenditure in Bahrain has been relatively volatile in recent years with sharp increases noted in 2006, 2010, and 2012 owing to the housing projects by the Ministries of Housing and Works. The expansion in expenditures was facilitated by a US$5 billion increase in public debt between 2008 and 2012.

The absence of corporate and personal income taxes as well as excise and sales taxes, creates a narrow tax base for the Maldives. The tax environment forces the government to rely heavily on tourism, property leasing and import tariffs for revenues. Despite minimal revenue sources, growth in revenues has steadily increased by 26 percent between 2007 and 2012 (see table 5.15). The growth in revenues was offset by a higher rate of growth in expenditures resulting in a deficit of 19 percent of GDP in 2012. The growth in expenditure, peaking at 61 percent of GDP in 2012, can be attributed to the public sector wage bill and social welfare payments amounting to 23 percent and 9 percent of GDP, respectively.[33] The wage bill rose by 134 percent, and subsidies increased threefold between 2007 and 2012.

As a result, the Maldives has had a persistent fiscal deficit reaching a high of 29 percent of GDP in 2009. Deficits have resulted in increases

Table 5.15 Maldives: Government operations

	2007	2008	2009	2010	2011	2012
Revenues	55.8	46.0	35.9	37.0	43.4	44.2
Tax Revenues	21.5	20.9	16.2	17.9	18.5	15.8
Import duties	15.5	15.2	11.0	9.7	8.9	8.9
Tourism tax	4.1	3.5	3.1	4.3	7.9	8.5
Non tax revenue	26.6	21.9	15.2	18.0	17.7	17.2
Royalties, land & resort rent	13.7	10.0	6.2	10.1	9.4	9.1
Grants	7.7	3.2	4.5	1.1	1.0	1.0
Expenditure and Net Lending	61.4	63.1	64.9	54.8	47.5	47.9
Current	48.5	46.3	51.9	45.2	41.0	41.5
Salaries and allowances	16.3	22.1	28.1	21.2	17.6	19.3
Social welfare	16.2	6.8	7.0	8.1	6.5	5.7
Capital	61.7	63.1	64.9	54.8	47.5	47.9
Net lending	−0.3	−1.0	−0.9	−1.3	−1.4	−1.1
Fiscal balance	**−5.6**	**−17.1**	**−29.0**	**−17.8**	**−4.1**	**−3.7**

Source(s): International Monetary Fund (2011).

in the debt-to-GDP ratio from 35 percent in 2007 to 81 percent in 2012, as an additional 18 percent of debt-to-GDP was accumulated in 2012 to close the deficit. Debt management in the Maldives remains extremely vulnerable to fiscal slippage that needs to be offset by debt. The government's thrust to create an evacuation fund and achieve carbon neutrality by 2019 requires significant fiscal consolidation to deter further increases in the deficit and debt similar to those incurred to the Asian Tsunami.

Mediterranean Sea

Although Cyprus' 2012 expenditure of 45 percent of GDP made it one of the most prudent spenders in the European Union, revenues of 40 percent of GDP ranked the government in the bottom half for revenue collection (see table 5.16).

The Cypriot government recorded deficits between 2004 and 2012, with the exception of the surpluses gained from a real estate boom in

Table 5.16 Cyprus: Government operations

	2004	2005	2006	2007	2008	2009	2010	2011	2012
(Percent of GDP)									
Revenue	38.8	40.9	41.7	44.6	43.1	40.1	40.9	39.9	39.6
Tax Revenue	32.6	26.7	27.9	32.2	30.6	26.4	26.5	26.4	25.7
Other Revenue	6.2	14.6	14.2	12.8	12.5	13.7	14.4	13.5	13.9
Expenditure	42.9	43.4	42.9	41.2	42.1	46.2	46.2	46.3	45.9
Current Expenditure	38.7	39.8	39.1	37.3	38.2	40.7	40.9	41.7	42.2
Interest Payments	3.3	3.6	3.3	3.0	2.8	2.6	2.2	2.4	3.1
Capital Expenditure	4.2	3.6	3.8	3.9	4.0	5.5	5.2	4.5	3.8
Fiscal Balance	-4.4	-2.1	-0.9	3.82	1.0	-6.1	-5.3	-6.4	-6.3
Public Debt									
(US millions)									
Total Public Debt	11,108	11,741	11,910	12,799	12,345	13,739	14,164	17,880	19,927
Domestic Debt	8,088	8,854	9,305	9,867	8,880	8,190	7,166	8,401	9,863
External Debt	3,019	2,887	2,605	2,932	3,465	5,549	6,998	9,479	10,064
(Percent of GDP)									
Total Public Debt	70.9	69.4	64.7	58.8	48.9	58.5	61.3	71.5	86.6
Domestic Debt	51.7	52.3	50.5	45.3	35.2	34.9	31.0	33.6	42.9
Foreign Debt	19.3	17.1	14.1	13.5	13.7	23.6	30.3	37.9	43.8

Source(s): Central Bank of Cyprus (2012) and International Monetary Fund (2014a, 2007).

2007 and 2008. The increase in fiscal deficits after the 2008–2009 crisis averaged 6 percent of GDP. This escalated concerns about the impact that the continued deficit would have on borrowing. Significant fiscal deterioration and the contraction of growth in the economy reversed the decline in public debt, which had reached a decade low of 49 percent of GDP in 2008. The mounting deficit between 2008 and 2012 resulted in an increase in public debt by 38 percent of GDP. This occurrence has also coincided with the reduction of sinking funds.[34]

Cyprus' exposure to the Eurozone Crisis was most extreme as the intricate economic relationship between Greece and Cyprus precipitated a debt fallout. As the demand for Cypriot exports to Greece—its main trading partner—fell and as the banking crisis evolved, debt ratios rose significantly. Attempts to stabilize the economy included a bilateral loan from Greece in 2012, a debt swap as well as a multilateral debt bailout. This resulted in a debt-to-GDP ratio of 86 percent in 2012 and a forecasted 108 percent in 2019 (International Monetary Fund 2014c).

Another contributor to public debt in 2012 was the Government's recapitalization of the financial sector during the Cypriot financial crisis, including a bond for shares bailout of the Cyprus Popular Bank. This increased budgetary financing to €7,261 million (Republic of Cyprus Ministry of Finance 2013). The worsening economic conditions have resulted in credit downgrades which have increased the cost of borrowing. Debt servicing obligations propelled debt levels to 112 percent of GDP in 2013 (International Monetary Fund 2014c). This makes debt servicing a primary focus for this economy as the impact of the credit downgrades serves to increase interest payments which could trigger a liquidity crisis.

South China Sea

Singapore has maintained a fiscal surplus over the past two decades, with the absence of net debt as assets far outweighing liabilities (see table 5.17). This poses a unique scenario for the AIMS region, as its most indebted country is the benchmark of regional and global fiscal discipline. Singapore has not had external debt since 1994 and public debt has only been accumulated through its bond market. The government acquires domestic debt through two instruments: Singapore Government Securities (SGS) for the development of the domestic debt market and the Special Singapore Government Securities (SSGS) which are non-tradable bonds that fulfill the investment needs of the national pension fund.[35] No government borrowing is used to offset government expenditure as the Protection of Reserves Framework in Singapore's Constitution and the Government Securities Act, prohibit the government from spending income raised from borrowing. Therefore, all debt raised from the SGS and SSGS is invested and there is no net debt as returns exceed the debt servicing costs.

Table 5.17 Singapore: General government operation

	1990–1999	2000–2004	2005	2006	2007	2008	2009	2010	2011	2012
(Singapore dollars, billions)										
Revenue	35.9	39.4	43.3	47.9	65.7	64.3	50.8	69.8	82.4	80.6
Expenditure	24.5	29.3	26.3	30.9	33.0	47.1	52.9	46.7	57.7	61.1
Fiscal Balance	11.4	10.2	16.9	17.0	32.7	17.3	-2.0	23.1	24.7	19.6
(Percent of GDP)										
Revenue	31.8	23.4	20.2	20.1	24.0	24.2	17.7	21.6	24.5	23.3
Expenditure	21.2	17.4	12.3	12.9	12.1	17.7	18.4	14.4	17.1	17.6
Fiscal Balance	10.6	6.0	7.9	7.1	12.0	6.5	−0.7	7.2	7.4	5.7
Public Debt										
Gross debt (Singapore, billions)	80.2	159.2	200.0	206.4	234.1	255.5	291.5	321.2	354.02	385.0
Gross debt (% of GDP)	72.0	93.2	93.5	86.4	85.6	96.3	101.5	99.3	105.18	111.0

Source(s): International Monetary Fund (2013e).

Factors Likely to Affect Debt Trends in the AIMS SIDS in the Near Future

Many of the AIMS SIDS face an impending challenge to debt management, as their debt levels have marginally surpassed the debt-to-GDP sustainability benchmark of 60 percent. Such sustainability is further threatened by a number of factors that are likely to affect regional debt trends in the future.

LDC Graduation

Three of the region's most economically fragile nations remain under LDC status (Comoros, Guinea Bissau, and Sao Tome). However, recent successes in debt reduction due mainly to debt relief and concessional funding have created the fiscal space for these economies to pursue development strategies. As these economies continue to grow, a baseline scenario of reduced aid must be contemplated, as the countries approach the threshold of being declassified as LDCs and losing access to debt relief and concessional funding.

The region has two recent examples of how debt can be affected with the graduation from LDC classification. These experiences should serve as lessons for future graduands. Following graduation from LDC status in 2007, Cabo Verde saw an increase in external aid, as it was given a five-year transitional window by development partners to strengthen economic independence. Being one of the world's highest per capita recipients of ODA, the continued inflow of ODA proved important to debt stabilization efforts, as a large share of the government's budget is covered by development assistance. The transitional period thus created the fiscal space for the island to maintain stable external and domestic debt figures. The lasting post-transition impact is, however, yet to be seen.

ODA is of lesser importance to the Maldives, as the debt pressures of graduation are more aligned with the loss of preferential treatment. For the Maldives, which has a significant amount of external debt, the loss of preferential markets will create short-term foreign exchange shortages. This is expected to increase the difficulty of repaying foreign currency denominated debt, which could deepen the already high risk of debt distress.

Vulnerability to Environmental Shocks

The environmental challenges of SIDS result from their small size, high susceptibility to natural hazards, low economic resilience and limited human and technological capacity for mitigating and adapting to the effects of climate change (Nurse and Sem 2001). AIMS SIDS are no different in this regard as they are amongst the most vulnerable to the potential hazards of climate change and the consequent rise in sea levels, despite their minute contribution to global greenhouse gas (GHG) emissions.[36] The very existence of low-lying islands such as the Maldives and Seychelles and coastal plains such as Guinea-Bissau, Mauritius, Bahrain, and Singapore means that more than half of AIMS SIDS are threatened by the sea level rise induced by climate change.[37]

Other related consequences of this phenomenon such as extreme weather conditions, and the acidification of oceans are expected to negatively impact the region's water resources, agriculture and fisheries, infrastructure and general economic development. The inadequacy of resources to mitigate losses associated with climate change risks threatens to increase debt levels in the region, as recovery programs related to climate change disasters have to be borne by these small states. This was a major concern at the regional preparatory meeting of AIMS SIDS, as the urgent need for development partners to replenish the Climate Adaptation Fund was noted. There was also an appeal from regional heads of state to fulfill pledges made in relation to the Green Climate Fund.[38]

Concluding Remarks

Debt accumulation and overhang are critical issues facing AIMS SIDS. With the onset of the recent global economic crisis and the region's extreme vulnerability to climate change, most AIMS SIDS have experienced a rapid accumulation in public debt and inflated levels of debt servicing.

Like the SIDS of the Caribbean and the Pacific, debt accumulation in AIMS SIDS has been largely influenced by international shocks and natural disasters. The global crisis worsened fiscal deficits in the region while disaster recovery efforts, especially in the islands of the Maldives and Seychelles, were partly funded by increases in debt. More importantly, however, domestic conditions, such as political uncertainty, have contributed to increasing debt levels among some AIMS SIDS. For instance, political instability in Guinea Bissau, Bahrain, and Comoros has deterred investments, suppressed growth and has precluded these countries from receiving the full benefit of debt relief (in the case of Guinea-Bissau and Comoros) or donor contributions.

Notwithstanding these challenges, the region has undertaken several initiatives to contain public debt within sustainable thresholds. Debt relief from the HIPC and the MDR Initiatives has helped to significantly reduce the external debt burdens of Comoros, Guinea-Bissau, and São Tomé and Principe. Debt restructuring efforts have also played an integral role in reducing the region's debt over time. Over the last decade, Cabo Verde, Comoros, Cyprus, Guinea-Bissau, Maldives, Seychelles, and São Tomé and Principe have all engaged in debt restructuring efforts with multilateral or bilateral creditors. Bahrain and Mauritius are examples of two countries in the region with sound debt management practices. Both countries have institutionalized a public debt ceiling of 60 percent and have implemented several policy measures to ensure that their debt trajectories remain within sustainable levels. Likewise, Seychelles has implemented several reform measures to better manage public finances to reduce debt accumulation.

Despite these efforts there are other initiatives that these countries as well as the international agencies such as the UN, IMF, and IDB should investigate and adopt in order to properly manage debt in the region. Given the extreme economic and environmental vulnerabilities of AIMS SIDS, development concessions must be extended to them. At the international level, this implies that the traditional measure of using income per capita to determine SIDS eligibility for aid/concessional financing needs to be revised to include more appropriate measures such as Economic and Environmental Vulnerability Indices.

Notes

* All authors contributed equally to this work.

1. While the United Nations Office of the High Representative for the Least Developed Countries, Landlocked Developing Countries and Small Island Developing States (UN-OHRLLS) declassified Cyprus as a Small Island Developing State (United Nations 2010), it will still be included in this analysis.
2. The Maldives is the smallest island in the region, occupying 298 square kilometers of land.
3. UN Office of the High Representative for the Least Developed Countries, Landlocked Developing Countries and Small Island Developing States
4. Seven of the AIMS SIDS have a medium to high level of development. All AIMS countries rank higher than their respective geographic regions—Africa, Asia, Europe, Arab States—except for Cyprus which is marginally led in development by Europe.
5. This was outlined in their 2008 Public Debt Management Act (PDMA), which essentially seeks to meet the borrowing needs of the government in a manner that minimizes the cost of the debt portfolio within an acceptable level of risk. See also Mauritius: 2013 Article IV Consultation, International Monetary Fund (2013c).
6. http://banksbahrain.org/news&key=feature_gulf_monetary_union_still_has_currency_despite_eurozone_crisis.
7. Comoros, Guinea-Bissau and Sao Tome and Principe were excluded due to comprehensive debt relief through the HIPC and MDR Initiatives. Seychelles' relatively high public debt trend between 2002 and 2008 represented an outlier and was therefore not included in the sample average.
8. Country Report Bahrain (Rabobank 2006). http://www.britishcouncil.org /ar/tvetarabia-bahrain-country-risk-research-06.pdf.
9. World Bank (2009).
10. More specifically, between 2001 and 2006, Guinea-Bissau, São Tomé and Principe, and Seychelles registered public debt ratios of 237 percent, 321 percent, and 170 percent, respectively. It should be noted that while Seychelles and São Tomé and Principe recorded such high level of indebtedness, their debt ratios were falling at a faster rate than any other AIMS SIDS.
11. UNDP (2009).
12. Including the introduction of a floating exchange rate in 2008 and by the end of 2009, the Seychelles Rupee stabilized at the market rate and inflation was reduced to single digits.
13. Due to political instability and capacity constraints, a full PRGF was only finalized in 2006.
14. ODA (as a percentage of GNI) received in the region has decreased from 13 percent in 2003 to 10 percent in 2008.
15. Data was taken from the World Bank (2013b). Data was unavailable for Bahrain, Cyprus, and Singapore.

16. The Government and the IMF agreed on a stand-by arrangement in February 2009. See International Monetary Fund (2010). Seychelles: Third Review under the Stand-By Arrangement (International Monetary Fund 2010).
17. International Monetary Fund (2013d). Regional Economic Outlook for Sub-Saharan Africa.
18. UNDP (2010c), Monthly Bulletin of Statistics, April.
19. UNDP (2013c), Official Development Assistance. http://www.undp.org/content/dam/undp/library/Poverty%20Reduction/Inclusive%20development/Towards%20Human%20Resilience/Towards_SustainingMDGProgress_Ch5.pdf.
20. Note that the high income, oil rich, country of Bahrain remains exempt from most multilateral debt relief agreements available to other AIMS SIDS; similarly, the income ranking of Singapore and Mauritius, when coupled with prudent macroeconomic policies, has eliminated the need to undertake any form of debt restructuring.
21. Debt Restructuring Initiatives Paper for the Commonwealth Secretariat (Robinson 2010).
22. Fitch rating agency raised Seychelles' rating from Issuer Default to B and assessed their outlook as positive.
23. The IMF purports that significant change in GDP growth can be decoupled from debt accumulation in the Maldives. The 2008 Debt Sustainability Analysis notes that with the exception of the 2004 tsunami and the following year, sensitivity tests project that the island's debt would fall through to 2013. IMF/World Bank (2009).
24. Singapore's gradual increase in public debt prior to the recession will be discussed in a latter section.
25. Despite the decline, Singapore was the only country in the region to record a positive trade balance.
26. The IMF categorizes HTBEs as countries where travel exports exceeded 10 percent of GDP for at least one year in 1998–2007 (International Monetary Fund 2008). The HTBEs in the region are: Cyprus, Singapore, Seychelles, Maldives, Mauritius, and Cabo Verde.
27. No remittance receipts were reported for Bahrain as it is considered a remittance sending country. However, given the country's wealth and natural endowment of fuel it is anticipated that the regional remittance receipts are unaffected by the absence of these figures.
28. The main remittance senders to the region are the United Kingdom, Portugal, France, Belgium, and the United States of America.
29. Guinea Bissau ranks 176 of 186 countries on the United Nations Human Development Index (UNDP 2013a).
30. Guinea Bissau had a cholera outbreak in 2012.
31. GFDRR—Global Facility for Disaster Reduction.
32. The increase has been driven by an expansionary fiscal policy in the aftermath of the tsunami. See International Monetary Fund (2011). Maldives Debt Sustainability Analysis Under the Debt Sustainability Framework for Low-income Countries.
33. A major part of the rise in subsidies is the absorption of increases in oil prices to stabilize electricity costs below market rates.

34. Sinking funds are interest-bearing accounts that were created and maintained by the Central Bank of Cyprus for the purpose of repaying specific government loans. In 2008, sinking funds were significantly reduced to honor debt obligations and were further depleted in subsequent years (Central Bank of Cyprus 2012).
35. Central Provident Fund (CPF).
36. Intergovernmental Panel on Climate Change (IPCC) (2001).
37. Maldives, among all SIDS, has the largest share, approximately 100 percent, of its population in a low elevation coastal zone (LECZ). Bahrain closely follows with roughly 80 percent of its population living in a LECZ. Note that LECZs are contiguous areas along the coast that is less than ten meters above sea level. See McGranahan, Balk and Anderson (2007).
38. Regional Preparatory Meeting of Small Island Developing States of the AIMS, July 17–19, 2013 Mahe Seychelles (UNDP 2013b).

References

Adams, R. H. J. 2006. "Remittances, Poverty, and Investment in Guatemala." In *International Migration, Remittances and the Brain Drain*, edited by Ç. Özden and M. Schiff. Washington, DC: The World Bank and Palgrave Macmillan.

African Development Bank (AfDB). 2014. *Socio-Economic Database*. Abidjan: African Development Bank.

Bahrain Economic and Development Board. *Kingdom of Bahrain Economic Yearbook 2013*. http://www.bahrainedb.com/en/EDBDocuments/Bahrain-Economic-Yearbook.pdf.

Central Intelligence Agency (CIA). 2013. *The World Factbook*. Washington, DC: Central Intelligence Agency. https://www.cia.gov/library/publications/the-world-factbook.

Centre for Research on Epidemiology of Disasters (CRED). 2013. *Emergency Events Database (EM-DAT): The International Disaster Database*. Brussels, Belgium: Universite` Catholique de Louvain. http://www.emdat.be/database.

Global Facility for Disaster Reduction and Recovery (GFDRR). 2013. *The Union of Comoros- Resilience Over Reaction to Natural Disasters*. Washington, DC: World Bank. https://www.gfdrr.org/node/1519.

Intergovernmental Panel on Climate Change (IPCC). 2001. *Climate Change 2001: Impacts, Adaptation and Vulnerability*. Geneva, Switzerland: Intergovernmental Panel on Climate Change (IPCC).

International Monetary Fund (IMF). 2007. *Cyprus: 2006 Article IV Consultation.: IMF Country Report No. 07/76*. Washington, DC: International Monetary Fund.

———. 2008. *Cape Verde: Selected Issues: IMF Country Report No. 08/243*. Washington, DC: International Monetary Fund. http://www.imf.org/external/pubs/ft/scr/2008/cr08243.pdf.

———. 2010. *Seychelles: Third Review Under the Stand-By Agreement*. Washington, DC: International Monetary Fund.

———. 2011. *Maldives: 2010 Article IV Consultation.* Washington, DC: International Monetary Fund. http://www.imf.org/external/pubs/ft/scr/2011/cr11293.pdf.

———. 2012a. *Cape Verde: Second Review Under the Policy Support Instrument and Request for Waivers of Nonobservance of Assessment Criteria.* Washington, DC: International Monetary Fund. http://www.imf.org/external/pubs/ft/scr/2012/cr1229.pdf.

———. 2012b. *Democratic Republic of São Tomé and Príncipe: Joint IMF/World Bank Debt Sustainability Analysis.* Washington, DC: International Monetary Fund. http://www.imf.org/external/pubs/ft/dsa/pdf/2012/dsacr12216.pdf.

———. 2013a. *Cyprus: Second Review Under the Extended Arrangement Under the Extended Fund Facility.* Washington, DC: International Monetary Fund. http://www.imf.org/external/pubs/ft/scr/2013/cr13374.pdf.

———. 2013b. *Guinea Bissau. 2013 Article IV Consultation- Debt Sustainability Analysis.* Washington, DC: International Monetary Fund. https://www.imf.org/external/pubs/ft/dsa/pdf/2013/dsacr13197.pdf.

———. 2013c. *Mauritius: 2013 Article IV Consultation.* Washington, DC: International Monetary Fund. http://www.imf.org/external/pubs/ft/scr/2013/cr1397.pdf.

———. 2013d. *Regional Economic Outlook for Sub-Saharan Africa: Building Momentum in a Multi-Speed World.* Washington, DC: International Monetary Fund.

———. 2013e. *World Economic Outlook Database.* Washington, DC: International Monetary Fund. http://www.imf.org/external/pubs/ft/weo/2013/02/weodata/index.aspx.

———. 2014a. Cyrpus: *Third Review Under the Extended Arrangement Under the Extended Loan Fund Facility. IMF Country Report No. 14/92.* Washington, DC: International Monetary Fund.

———. 2014b. *Low Income Countries Debt Sustainability Analysis for Poverty Reduction and Growth Trust (PRGT)-Eligible Countries, May 1, 2014.* Washington, DC: International Monetary Fund.

———. 2014c. World Economic Outlook. Washington, DC: International Monetary Fund.

International Monetary Fund and World Bank (IMF/World Bank). 2009. *Maldives: Joint IMF/World Bank Debt Sustainability Analysis Under the Debt Sustainability Framework for Low Income Countries.* Washington, DC: International Monetary Fund.

International Trade Centre (ITC). 2013. *Trade Statistics.* Geneva: International Trade Centre. www.intracen.org.

Maldives Monetary Authority. 2013. *Research and Statistics.* http://www.mma.gov.mv/statis.php.

McGranahan, Gordon, Deborah Balk, and Bridget Anderson. 2007. "The Rising Tide: Assessing the Risks of Climate Change and Human Settlements in Low Elevation Coastal Zone." *Environment and Urbanization* 19(1): 17–37.

Nurse, L. and Graham Sem. 2001. "Small Island States." In *Climate Change 2001: Impacts, Adaptation, and Vulnerability*, edited by J. J. McCarthy, et al., 842–975. Cambridge: Cambridge University Press.

Organisation for Economic Co-operation and Development (OECD). 2011. "Socio-Economic Context and Impact of the 2011 Events in the Middle East and North Africa Region." *MENA-OECD Investment Programme.* Paris: Organisation for Economic Co-operation and Development.

Rabobank. 2006. *Country Report Bahrain.* Utretch, The Netherlands: Rabobank Group.

Republic of Cyprus Ministry of Finance. 2013. *Annual Report: Public Debt Management. Nicosia: Ministry of Finance.* http://www.mof.gov.cy/mof/pdmo/pdmo.nsf/0/FF9F9CCE6F27A2DCC225782B002F21CC/$file/Annual%20Report%202012.pdf

Robinson, Michele. 2010. "Debt Restructuring Initiatives Paper." Prepared for the Commonwealth Secretariat Small States Biennial Conference, London, July 28–29.

Standard and Poor's Rating Services. 2014. *Sovereign Ratings List.* Retrieved on June 8, 2014 from: http://www.standardandpoors.com/ratings/sovereigns/ratings-list/en/us/?sectorName=Governments.

United Nations. 2010. *AIMS Regional Synthesis Report for the Five-Year Review of the Mauritius Strategy for Further Implemetation of the Barbados Programme of Action for Sustainable Development of SIDS (MSI+5).* New York: United Nations. http://www.unohrlls.org/UserFiles/File/AIMS_Regional_Synthesis-MSI5-Final.pdf.

United Nations Development Programme (UNDP). 2009. *Assessment of Devlopment Results: Evaluation of UNDP—Seychelles.* New York: United Nations Development Programme. http://web.undp.org/evaluation/documents/ADR/ADR_Reports/Seychelles/ExecSum-Seychelles-en.pdf.

———. 2010a. *Achieving Debt Sustainability and MDGs in Small Island Developing States.* New York: United Nations Development Programme.

———. 2010b. *Achieving Debt Sustainability and MDGs in Small Island Developing States: The Case of the Maldives.* New York: United Nations Development Programme. http://www.undp.org/content/dam/aplaws/publication/en/publications/poverty-reduction/poverty-website/achieving-debt-sustainability-and-the-mdgs-in-sids-the-case-of-the-maldives/Maldives_Debt_Sustainability_MDGs.pdf.

———. 2010c. *Monthly Statistical Bulletins.* New York: United Nations Development Programme.

———. 2013a. *Human Development Index.* New York: United Nations Development Programme. https://data.undp.org/dataset/Table-1-Human-Development-Index-and-its-components/wxub-qc5k.

———. 2013b. *Regional Preparatory Meeting of Small Island Developing States of the AIMS: Outcome Document*, Victoria, Seychelles, July 17–19.

———. 2013c. "Official Development Assistance." In *Towards Human Resilience: Sustaining MDG Progress in an Age of Economic Uncertainty.* New York: United Nations Development Programme.

UN-OHRLLS. 2013. *Small Island Developing States (SIDS) Statistics.* United Nations Office of the High Representative for the Least Developed Countries.

World Bank. 2009. *Global Economic Prospects.* Washington, DC: World Bank.

———. 2013a. *Migration and Remittances Database.* Washington, DC: The World Bank. http://econ.worldbank.org/WBSITE/EXTERNAL/EXTDEC/EXTDECPROSPECTS/0,,contentMDK:22759429~pagePK:64165401~piPK:64165026~theSitePK:476883,00.html.
———. 2013b. *World Development Indicators Database.* Washington, DC: World Bank. http://databank.worldbank.org/data/views/variableSelection/selectvariables.aspx?source=world-development-indicators.
———. 2014. *Guinea Bissau Overview.* Washington, DC: World Bank. http://www.worldbank.org/en/country/guineabissau/overview#1

Chapter 6

Prospects for the Growth of Debt in Selected SIDS*

David F. Tennant and Kario-Paul Brown

The previous chapters have shown that numerous Small Island Developing States (SIDS) currently have public debt to GDP ratios in excess of 60 percent (the broadly accepted threshold for sustainable levels of public debt). A significant proportion of these countries, particularly those in the Caribbean, are now registering public debt to GDP levels of more than 100 percent. In spite of these high levels of indebtedness, only five SIDS (Comoros, Guinea-Bissau, Guyana, Haiti, and Sao Tome and Principe) have been considered poor enough or indebted enough to benefit from debt-relief schemes such as the Heavily Indebted Poor Countries (HIPC) initiative and the Multilateral Debt Relief Initiative (MDRI).[1]

Many have thus come to the conclusion that "the problem of indebtedness in other small island developing states has been largely disregarded by the international policy community."[2] This then leaves SIDS to independently resolve their indebtedness problems mainly by "cutting public expenditures, raising taxes and seeking debt restructuring with individual creditors on an ad-hoc basis" (Hurley 2011). For the most severely indebted SIDS these efforts have not been adequate, as many continue to have massive debt overhangs.

This is critical, because in a context where public debt burdens severely limit fiscal space and constrain countries' ability to pursue growth-enhancing and poverty-reducing policies, the need for external financing is urgent. This leaves highly indebted SIDS in a precarious position, as the IMF and World Bank have noted that they consider debt sustainability assessments when determining whether countries will be given access to financing.[3] If the assessments are not favorable, then these countries are forced into commercial borrowing relationships, with rates and terms that are likely to exacerbate their debt levels.

While low to moderate levels of debt can be beneficial to development, high debt levels bring into question the issue of sustainability, as they increase the probability of default. More fundamentally, proponents of the debt overhang hypothesis argue that high levels of debt have adverse effects on economic growth. Caner, Grenner, and Koehler-Geib (2010), for example, find that countries with a public debt to GDP ratio of above 77 percent experience adverse growth effects. Reinhart and Rogoff (2010) estimate a slowdown in economic growth when the debt to GDP ratio hits 90 percent.[4]

More recently, however, Pescatori, Sandri, and Simons (2014) found evidence indicating that while high levels of debt may be associated with fiscal and monetary policy actions that destabilize economic growth, the debt trajectory of a country can be even more important in understanding future growth prospects. They found that countries with high but declining levels of debt have grown just as fast as their peers.

So where does this leave SIDS? Many have very high levels of debt, but where do the trajectories of their debt suggest they will end up in the medium term, and what are the implications for sustainability and growth? This chapter uses an adaptation of the IMF's debt sustainability framework to assess the debt trajectories of the SIDS with some of the highest debt levels. We focus on the Caribbean region, as the previous chapters have indicated that this is where the largest concentration of highly indebted SIDS are located. The countries included in the study are Antigua and Barbuda, Belize, Barbados, Dominica, Grenada, Jamaica, and St. Kitts and Nevis. We examine debt sustainability in these countries in a consistent macroeconomic framework, to ascertain whether additional policy actions are needed to prevent debt default in the medium term (i.e. up to 2020). The projection exercises are complemented with scenario analyses that assess the vulnerability of projected medium-term outcomes to macroeconomic shocks.

The chapter will proceed by briefly outlining the conceptual framework and guiding principles that have been used in the analysis, summarizing the methodology used, describing the scenarios to be assessed, and presenting the results of the scenario and sensitivity analyses. The concluding section will seek to draw general lessons for highly indebted SIDS.

Conceptual Framework and Guiding Principles

The analyses in this chapter focus on the impact of fiscal policy measures and macroeconomic shocks on debt sustainability. The IMF (2002, 4) defines debt sustainability as a situation in which a "borrower is expected to be able to continue servicing its debts without an unrealistically large future correction to the balance of income and expenditure."

So for the purposes of this study, debt is unsustainable when: a debt restructuring is already needed or is highly likely to be needed; the borrower accumulates debt faster than the growth in its capacity to service the debt; or the borrower lives beyond its means, that is by accumulating debt with the knowledge that servicing the debt will require a major retrenchment.[5]

Assessing debt sustainability requires, inter alia, projecting for a country its flows of revenues and expenditures. Fiscal policy sustainability and public debt sustainability are clearly interrelated concepts. A country's fiscal policy stance can be regarded as unsustainable if, in the absence of an adjustment, the government would eventually be unable to service its debt.[6] The IMF thus typically assesses a country's current fiscal policy by judging whether the existing fiscal balance is consistent with a stable debt to GDP ratio. The difference between the existing primary fiscal balance and the debt-stabilizing primary fiscal surplus is used as an indication of how much effort is required to achieve a stable debt ratio (IMF 2011d).

Assessing the debt trajectory of a country requires a proper analysis of fiscal policy, which must be based on a solid foundation of realistic macroeconomic baseline assumptions. Assumptions must be made about, inter alia, rates of economic growth, interest rates on public debt, and the likelihood that fiscal risks might arise.[7] Given the currency denomination of the debt, projections of exchange rate changes are also important, as are projections of financial market developments. Projections also frequently "incorporate judgments, based on historical and cross-country experience, of what adjustment is politically and socially feasible."[8] The IMF (2011d, 6) further notes that "it is equally important to stress test the underlying assessment of both fiscal and debt sustainability with respect to deviations from baseline assumptions for all these variables. Higher interest rates...or lower growth assumptions could, for example, result in less favorable debt dynamics, requiring an increase in the primary balance needed to stabilize the debt ratio."

Methodology

The principles outlined above have all been adhered to in the analyses conducted in this chapter. The IMF's (2002) framework for debt sustainability analysis is based on a two-pronged approach.[9] In analyzing the public debt, the trajectory of domestic currency denominated debt and foreign currency denominated debt are demarcated to reflect exchange rate dynamics and different levels of credit risk. The evolution of the budget deficit is examined to assess the sustainability of the public debt, while the evolution of the current account is examined for the external debt (which includes public and private external debt).

This study adopts the basic approach of the IMF, but modifies the methodology to reflect certain peculiarities of the debt structure of the Caribbean economies being studied. For example, in many of the countries, the level of private external debt is miniscule relative to the size of total external debt. Additionally, information on private external debt is usually very difficult to source. Debt sustainability in this study is therefore assessed by way of the dynamics of the central government budget rather than the two-pronged approach that includes the balance of payments (BOP). In the countries where the size of the public external debt creates significant balance of payment pressures, a bird's eye assessment of balance of payment prospects is conducted. For most of the selected countries, the evolution of public debt is influenced heavily by operations of the Central Government. Within this context and for comparability purposes, our study focuses on the dynamic evolution of the central government's debt. Historical data show that multilateral and bilateral debt are usually large relative to the total debt of the selected countries. These sources of debt usually have discounted interest rates and favorable amortization profiles relative to commercial debt.

Additionally, while off-budget events such as contingent liabilities were a major contributor to the accumulation of debt in the Caribbean, the absorption by central government of future such liabilities was not included in the extrapolations below because they are entirely exogenous to the dynamics of fiscal deficits and debt service. So whatever debt outcome is projected, an assumption about the amount of future realized contingent liabilities can simply be added. This chapter's focus, instead, is on the effect of different expectations about fiscal performance and interest rate paths on the evolution of the debt stock. By doing so, we are able to draw conclusions about the degree of fiscal consolidation required for a sustainable debt profile. The algebraic derivation of the model used to conduct the scenario analysis is provided in an appendix.[10]

Scenario Descriptions

Projected debt/GDP ratios are computed for each of the countries examined in this chapter, and are compared across the scenarios articulated below.

1. In the *baseline scenario* we seek to estimate the status quo by assuming that the respective governments adhere to the fiscal policies that are currently being implemented or that are likely to be implemented based on publicly announced plans. In all the

countries examined, some corrective fiscal policies were already adopted or are in the pipeline, and, on that basis, improved debt/GDP ratios are projected. We assess whether those improvements are sufficient to place the countries on a sustainable debt trajectory, by comparing the computed debt/GDP ratios at the end of the projection period with an estimated debt default threshold. The default thresholds used for countries that have previously defaulted are the debt/GDP ratios existing at the time of their most recent default. For countries that have not previously defaulted, the default thresholds are computed as the debt/GDP ratios consistent with debt servicing costs that are greater than the primary balances plus an estimate of market support.[11]

2. A path of *least resistance scenario* was also created to project what would have happened to each country's debt/GDP ratio if the governments had done and continue to do nothing about their deficit positions. This scenario assumes that due to political exigencies or social pressures the respective governments rollback the corrective measures now being implemented and revert to the path that resulted in the currently high debt positions. The projected debt/GDP ratios are compared with the debt default threshold and highlight the increased likelihood of default. This scenario is useful as it highlights the dangers associated with fiscal profligacy and/or indiscipline.

The results of the above two scenarios are supplemented by additional analyses in which the focus is shifted away from projecting debt/GDP ratios to ascertaining the fiscal policy stances that will lead to two predetermined debt/GDP outcomes during the projection period. These two supplemental exercises are outlined below:

3. The *default scenario* is computed to ascertain the primary balance that is likely to cause each country to just cross the debt default threshold in any year in the projection period. It highlights, for the respective governments, the minimum primary surplus or maximum primary deficit that they need to maintain in order to avoid defaulting on their debt between now and 2020.
4. The *target scenario* focuses on the opposite end of the range by indicating the primary balance that is required to get countries to the targeted debt/GDP ratio of 60 percent within the projection period. It thus shows what is required to get each country to a debt level consistent with strong economic growth by 2020.

Assumptions and Results

The Baseline Scenario

The macroeconomic and policy assumptions and results of the baseline scenario are presented below. The sensitivity of the baseline results to macroeconomic shocks is also analyzed. This section assumes that readers have some familiarity with the countries being discussed.[12] To get the requisite country backgrounds, a prior reading of chapter 3 is encouraged.

Assumptions

The macroeconomic assumptions represent the path that inflation, real GDP, the exchange rate and the interest rates are expected to take over the forecast horizon. These variables are expected to have the most significant impact on the trajectory of the government's debt. To project the paths of these variables we use the average forecasts from Standard and Poor's Rating Services, Business Monitor International, Caribbean Information and Credit Rating Services, and the International Monetary Fund. The macroeconomic assumptions are presented in table 6.1.

The baseline scenario seeks to identify policies that have been implemented or are likely to be implemented based on the practicality of communicated medium-term fiscal policies. Budget presentations, medium-term fiscal policy documents (where available) and historical data provide the pool of information used to formulate the main policy assumptions. The assumptions for each country are outlined below.

Antigua and Barbuda

The government-imposed wage freeze expires in fiscal year 2012. Expenditure on programs grows at the rate of inflation in 2012, but for 2013 onward it is assumed that the wage bill increases by 2 percent above the inflation rate. Public sector transformation and pension reform yields incremental benefits over the long run. Capital expenditure is increased gradually from 2.4 percent of GDP in 2012 to 5 percent of GDP by 2020. As economic growth accelerates and fiscal slippage declines, the revenue enhancing measures already implemented lead to a gradual increase in the tax revenue-to-GDP ratio from 20.7 percent in 2012 to 23.5 percent by 2020. The overall primary balance strengthens from 1.5 percent of GDP in 2012 to 4.1 percent by 2020.

The negotiations to reschedule and normalize the 15 percent of Antiguan debt that is in arrears are concluded by the middle of 2012. It is assumed that accumulated credit due to suppliers is reduced over the medium term. The negative investor sentiments following the debt restructuring exercise, and the fragility of the international capital

Table 6.1 Baseline scenario

Countries	Baseline Debt-to-GDP Trajectory									Default Threshold	Baseline Assumptions*			
	2012	2013	2014	2015	2016	2017	2018	2019	2020	≥ Debt Ratio	Exchange Rate**	Interest Rate	Real GDP	Primary Balance
Antigua and Barbuda	75.0	71.7	67.6	63.9	59.9	55.3	49.9	44.3	38.9	77.0	0.0	3.8	3.8	2.8
Belize	78.6	78.5	77.7	75.8	73.7	71.0	67.8	64.9	62.4	91.0	0.0	6.5	2.4	3.0
Barbados	97.1	96.8	94.0	93.0	91.6	90.4	89.8	88.1	84.6	122.0	0.0	6.7	2.6	2.7
Grenada	87.0	91.2	94.6	98.9	101.0	102.9	105.8	108.7	111.9	76.6	0.0	4.6	2.5	−3.4
St Kitts and Nevis	85.5	80.9	78.0	74.4	69.7	64.0	58.0	51.4	45.0	113.0	0.0	2.7	2.1	3.4
Jamaica	126.6	124.4	123.0	120.8	119.9	119.8	120.0	118.5	115.2	130.0	5.0	8.0	1.4	4.1
Dominica	75.8	81.5	82.9	82.9	82.7	81.0	78.5	75.6	72.2	82.0	0.0	2.8	1.4	−0.9

Notes: *In absolute percentages except for the primary balance which is in per cent of GDP.
**Percent depreciation for each country's currency.

Source(s): Authors' Estimates

markets force the government to finance approximately 85 percent of its fiscal deficit through multilateral and bilateral debt. Short-term instruments issued to the Eastern Caribbean Central Bank (ECCB) and on the Regional Government Securities Market (RGSM) remain a source of financing, albeit small, as these creditors were excluded from the debt restructuring exercise that began in 2011.

Belize

It is assumed that no additional fiscal austerity measures are implemented. The tax revenue-to-GDP ratio remains largely unchanged at 23.3 percent throughout the forecast horizon. Wages and salaries grow at the rate of inflation, and capital expenditure remains largely unchanged at 4.2 percent. Programmed expenditure grows at 2 percent above the rate of inflation. The primary balance remains in a deficit position, gradually widening from 2.8 percent of GDP to 3.2 percent of GDP. Just over 83 percent of the fiscal deficit is financed from external sources, of which bilateral and multilateral borrowings account for close to 50 percent.

Barbados

It is assumed that some of the fiscal measures outlined in the nation's medium-term fiscal strategy (MTFS) document are implemented,[13] without any additional revenue measures over the forecast horizon. Tax revenues increase gradually from 26.7 percent of GDP in 2012 to 27.1 percent of GDP by 2020 as fiscal slippage declines. Wages and salaries remain constant in 2012 and 2013 but increase at the rate of inflation from 2014 onward. The recent divestment of a public body, public sector transformation, and pension reform yield incremental benefits over the long run. Given the fiscal policy initiatives, the primary balance strengthens from 1.2 percent of GDP in fiscal year 2012 to 3.8 percent of GDP by fiscal year 2020. The government finances 70 percent of its fiscal deficit mainly through domestic debt, of which 82 percent are fixed rate instruments. Approximately 30 percent of the total deficit is financed by external debt, mainly through multilateral and bilateral loans.

Grenada

A minimal number of austerity measures are implemented. The tax revenue-to-GDP ratio fluctuates mildly around a central tendency of 17.7 percent of GDP through to 2020. Wages and salaries grow by approximately 2 percent above the inflation rate, while expenditure on programs grows by approximately 2.5 percent above the inflation rate. Fiscal authorities gradually reduce capital expenditure to 8.9 percent of GDP over the forecast horizon, from a budgeted 10.2 percent in 2012. Given the fiscal policy stance, the primary balance gradually decreases from 4.4 percent of GDP in 2012 to almost 2.5 percent of GDP by 2020. Approximately 80 percent of the fiscal deficit is financed through external sources, of which bilateral and multilateral loans account for over 90 percent.

St. Kitts and Nevis

As economic growth accelerates, fiscal slippage is assumed to gradually decline leading to a modest increase in the tax revenue-to-GDP ratio. Tax revenues increase from 21.3 percent of GDP in fiscal year 2012 to approximately 24.2 percent of GDP in fiscal year 2020. The wage bill remains constant in fiscal year 2012 and 2013, thereafter it increases by 2 percent above the rate of inflation. Expenditure on programs grows at one percentage point over the inflation rate. Public sector transformation and pension reform yields incremental benefits over the long run. Capital expenditure gradually increases from 5.9 percent of GDP to 7.5 percent of GDP by fiscal year 2020. The fiscal policy initiatives lead to a primary balance that ranges between 2.4 to 3.6 percent of GDP over the forecast horizon.

The baseline scenario for St. Kitts and Nevis also assumes that there was 100 percent participation in their recently implemented debt exchange program (see chapter 3).[14] In this program, two-thirds of the debt holders opted for a 50 percent reduction in principal, and the remaining one-third opted for no reduction in principal. The 50 percent discount bonds are based on a monthly mortgage-style repayment structure with no grace period on principal. The discount bonds have a final maturity of 20 years, with the last payment due in March 2032. The coupon on these instruments is 6 percent for the first four years, dropping to 3 percent from March 2016 onward. The par bonds have a final maturity of 45 years and are also based on monthly mortgage-style payments. The coupon on the par bonds is 1.5 percent, with a grace period of 15 years applying to principal payments. Fiscal authorities complete the second stage of the debt exchange program by the third quarter of 2012. The second stage of the debt exchange program is a debt-for-land swap where all secured debt holders exchange bonds for land through a special purpose entity.

Jamaica

The assumption for Jamaica is that fiscal authorities neither roll back existing revenue enhancing measures nor implement any additional revenue enhancing measures. Clarendon Alumina Partners (CAP) is divested by the end of calendar year 2012. Tax revenue as a percentage of GDP fluctuates mildly around an average of 22.8 percent. The wage bill increases at the rate of inflation, while expenditure on programs increases at 3 percent above the inflation rate. Pension reform and public sector reform takes place over the long run but does not significantly impact the fiscal numbers over the forecast horizon. Capital expenditure remains largely unchanged at approximately 3.9 percent of GDP through to 2020. The fiscal policy stance leads to a primary balance that fluctuates mildly around an annual average 4.1 percent of GDP.

The exchange rate depreciates by approximately 4.5 percent annually. Just over 51 percent of the fiscal deficit is financed through domestic sources, of which variable rate or short-term fixed rate debt instruments account for approximately 52 percent. External private capital market support remains constant for the first two fiscal years but strengthens from the fourth year onward. Multilateral and bilateral debt accounts for over 80 percent of total deficit financing from external sources, declining to approximately 40 percent from the fourth year onward.

Dominica

For Dominica we also assume that policy makers neither roll back existing fiscal austerity measures nor implement any additional revenue enhancing measures. Tax revenues remain largely unchanged at 24.3 percent of GDP, and wages and salaries increase at the rate of inflation. Expenditure on programs increases at one percentage point above inflation. It is further assumed that 70.3 percent of the fiscal deficit is financed from external sources, of which multilateral and bilateral debt account for 90 percent.

Results

The projections under the baseline scenario are presented in table 6.1. Under this scenario only two of the countries studied are likely to default on their debt if additional policy measures are not implemented in the near term. Dominica and Grenada have a high probability of default, as projected debt/GDP ratios exceed the debt default threshold at some point over the forecast horizon. In the case of Grenada, the projections suggest that the debt default threshold will be exceeded from as early as 2012, indicating that a default is imminent if nothing further is done about the country's debt position. For Dominica, the default threshold is exceeded by 2014, which suggests that the central government has at most one to two years to implement the necessary remedial measures.

Although the level of debt for Jamaica and Belize are projected to be below the debt default threshold over the forecast horizon, high refinance risks exist due to fragile market confidence and significant debt maturities coming due within the projection period. In the case of Jamaica, over 78 percent of total domestic debt (US$8 billion), and over 50 percent of total external debt (US$4.4 billion) matures over the next ten years. Belize's weak credibility in international capital markets makes it challenging to raise the necessary foreign currency resources to fund foreign currency liabilities when they come due. Bond yields on Belize's US dollar denominated super bond have moved into the high teens. There is a risk that stress points can lead to a default scenario from 2019 onward when the super bond, which is 2.5 times larger than projected Net International Reserves (NIR), begins amortizing.

The projections under the baseline scenario thus suggest that only three of the seven countries studied (Antigua and Barbuda, Barbados, and St. Kitts and Nevis) should be satisfied with the fiscal measures that they have already implemented or are planning to implement. That satisfaction should, however, be tempered by the knowledge that these projections do not include contingent liabilities, historically the primary source of new debt in the region. Additionally, even for these countries, the sensitivity analysis below indicates that certain macroeconomic shocks can significantly increase their vulnerability.

Sensitivity Analysis

The sensitivity analysis examines the responsiveness of baseline fiscal indicators to shocks to baseline assumptions on key macroeconomic variables, such as real GDP, nominal interest rates and the exchange rate. For comparative purposes, the size of each shock is standardized, except in cases where there are one-off shocks. The results of the sensitivity analysis are presented in table 6.2 and are discussed below.

Table 6.2 Sensitivity analysis*

Countries	Exchange Rate Shock**		Interest Rate Shock***		Real GDP Shock/2	
	Debt-to-GDP	Chg	Debt-to-GDP	Chg	Debt-to-GDP	Chg
Antigua and Barbuda	46	7	41	2	95	56
Belize	86	24	67	5	84	22
Barbados	94	9	92	7	107	22
Grenada	131	19	116	4	166	54
St Kitts and Nevis	44	6	41	3	74	36
Jamaica	128	13	125	10	137	22
Dominica	85	13	74	2	100	28

Notes: *Fiscal year 2020 debt-to-GDP ratios.

**One-off shock of 30 percent depreciation for all countries except Jamaica where a permanent 4 percent is applied.

***0.5 Standard Deviation Shock.

Source(s): Authors' estimates.

A Permanent Shock to Real GDP Growth

For the countries that were most likely to default under the baseline scenario (Dominica and Grenada), a half of a standard deviation permanent shock to baseline real GDP growth causes a significant worsening of the debt/GDP ratio and increases the likelihood of default. By 2020 the debt/GDP ratios of Dominica and Grenada are projected to increase by 28 percent and 54 percent, respectively. The fragility of the Jamaican debt situation was highlighted in the baseline results and is confirmed here. This shock to GDP growth is projected to raise Jamaican debt levels by 22 percent of GDP by 2020, and thereby pushes the country over its debt-default threshold. Antigua and Barbuda is also particularly vulnerable to this shock, as it was not projected to default under the initial baseline assumptions, but is projected to cross the debt-default threshold by 2020 with the specified shock to GDP growth. A half standard deviation permanent shock to baseline real GDP growth raises Antiguan debt levels by approximately 56 percent of GDP. As was the case for Jamaica, a similar shock to Barbados' and Belize's baseline real GDP growth raises debt levels in fiscal year 2020 by approximately 22 percent of GDP. This is the mildest response to this shock to baseline real GDP growth, and for these countries do not lead to projected defaults. So the largest responses were recorded by Antigua and Barbuda and Grenada (56 and 54 percent, respectively, of additional debt), followed by St. Kitts and Nevis (36 percent).

A Permanent Shock to Interest Rates

This shock consists of a 250 basis point increase in the interest rate on commercial debt. The susceptibility of the Grenadian government to default on its debt is also highlighted by this hypothetical event, as it is once again projected to cross the debt-default threshold by 2020 when this shock is introduced. For Grenada, this threshold is crossed even though the shock results in a relatively small change to the debt/GDP ratio of 4 percent. Projected changes in the debt/GDP ratio of 2 percent for Antigua and Barbuda and Dominica, and 3 percent and 5 percent for St. Kitts and Nevis and Belize, respectively, do not similarly precipitate a crossing of the threshold. Even the countries which were most responsive to changes in the interest rates (Jamaica and Barbados with increased debt/GDP ratios of 10 percent and 7 percent, respectively) were not projected to default as a result of this shock.

Exchange Rate Shock

The size of the exchange rate shocks applied to the selected countries depends on the exchange rate regime of the countries examined. For the six countries that have a fixed exchange rate regime, a one-off 30 percent depreciation is applied to capture the possibility of a stressed exchange rate adjustment in times of severe balance of payment pressures. For Jamaica, which has a flexible exchange rate regime, a permanent shock is applied.

A 30 percent depreciation in the EC dollar pushes both Dominica and Grenada over their debt-default thresholds by 2020, with increases in the debt/GDP ratios of 13 percent and 19 percent, respectively. Although the exchange rate shock is not projected to cause the Jamaican government to cross its default threshold, it significantly increases its vulnerability by pushing the country's debt/GDP ratio to within two percentage points of the threshold. A ¼ standard deviation shock to the rate of depreciation of the Jamaican dollar results in an increase in debt by 13 percent of GDP. Belize is most responsive to the exchange rate shock, with a one-time 30 percent depreciation in the Belizean dollar resulting in a 24 percent of GDP increase in debt levels by fiscal year 2020. By contrast, St. Kitts and Nevis, Antigua and Barbuda, and Barbados are least impacted by the exchange rate shock. A 30 percent depreciation of the EC and Barbadian dollars increases the debt/GDP ratios of these countries by 6 percent, 7 percent, and 9 percent, respectively, and does not dramatically heighten the likelihood of default.

Special Case Shock: St. Kitts and Nevis

In the baseline scenario for St. Kitts and Nevis it was assumed that a debt-for-land swap would be successfully implemented. If this assumption does not hold, and there is no debt-for-land swap, the simulations show that (ceteris paribus) the 2020 debt/GDP ratio for this country could increase by 27 percentage points over that which was initially projected in the baseline scenario. The debt-for-land swap is thus important, as the amount of secured debt is approximately 25 percent of total central government debt, and all secured debt holders are eligible to participate in the swap.

Least Resistance Scenario

The path-of-least-resistance scenario examines the trajectory of the debt/GDP ratio if fiscal authorities roll-back the fiscal austerity and other remedial measures that were implemented post-recession, with the aim of strengthening political appeal in the pre-election period. The macroeconomic variables stay the same as in the baseline scenario. The results of this scenario are presented in table 6.3.

Under the path of least resistance scenario, the debt/GDP ratios of all the countries examined, except for Belize, breach the debt default threshold at some point during the forecast horizon. With the reversal of the fiscal consolidation measures, the primary balance will deteriorate

Table 6.3 Path of least resistance scenario

Countries	Debt-to-GDP Trajectory									Default Threshold	Baseline Assumptions*			
	2012	2013	2014	2015	2016	2017	2018	2019	2020	≥ Debt Ratio	Exchange Rate**	Interest Rate	Real GDP	Primary Balance
Antigua and Barbuda	78.4	78.3	77.3	76.7	75.5	73.5	70.5	67.2	64.1	77.0	0.0	3.8	3.8	−0.8
Belize	81.4	84.3	86.7	87.9	89.0	89.4	89.1	89.3	89.9	91.0	0.0	6.5	2.4	0.5
Barbados	102.8	108.7	111.8	117.2	122.1	127.6	134.3	139.5	142.3	122.0	0.0	6.7	2.6	−2.8
Grenada	88.2	93.4	97.9	103.4	106.5	109.5	113.4	117.4	121.7	76.6	0.0	4.6	2.5	−4.5
St Kitts and Nevis	119.7	125.2	132.8	140.1	145.7	149.9	153.2	154.9	156.4	113.0	0.0	2.7	2.1	−4.9
Jamaica	130.0	131.0	133.1	134.4	137.2	141.0	145.5	148.1	148.6	130.0	5.0	8.0	1.4	2.1
Dominica	89.8	99.8	106.1	111.4	117.4	122.4	126.7	130.8	134.6	82.0	0.0	2.8	1.4	−4.6

Notes: *In absolute percentages except for the primary balance which is in percent of GDP.
**Percent depreciation for each country's currency.

Source(s): Authors' estimates.

for all countries, and will set off a chain reaction that leads to a substantial increase in debt. The rolling back of the fiscal measures leads to the most severe worsening of the primary balance (as a percent of GDP) in St. Kitts and Nevis (with an 8.3 percent reduction in the primary balance to GDP ratio), Barbados (5.5 percent), Dominica (3.7 percent), and Antigua and Barbuda (3.6 percent). In this scenario, all of these countries, with the exception of Barbados, are projected to cross the debt default threshold from as early as 2012. Barbados crosses her threshold in 2016, and the debt/GDP ratio is projected to continue to rise rapidly thereafter. Although Grenada and Jamaica experience the least severe worsening of their primary balances in the path of least resistance scenario (with reductions in the primary balance to GDP ratio of 1.1 percent and 2 percent, respectively), they are also projected to equal or cross their default thresholds from as early as 2012. In both countries the debt/GDP ratios are projected to continue to increase through to 2020.

Default Scenario

The default scenario highlights the fiscal policy stance that leads to a breach of the debt default threshold over the forecast horizon. The baseline forecasts for the macroeconomic variables are used. The results are presented in table 6.4.

From the baseline scenario it was already found that Dominica and Grenada are likely to default at some time before 2020 on the basis of their current fiscal policies. For the other countries, a comparison of the primary balance to GDP ratio in the default scenario, with that of the baseline scenario, gives an indication of the fiscal space that is available. For example, with a primary balance to GDP ratio of 3.2 percent in the default scenario and a marginally higher ratio of 3.4 percent in the baseline scenario, St. Kitts and Nevis has very little scope for expanding their primary balance. Even a small deviation from its current fiscal path is likely to lead to a default at some time over the projection period. Belize, Antigua and Barbuda, and Jamaica have a bit more fiscal room to maneuver, with differences between the baseline and default scenario primary balance to GDP ratios of 2.6, 2 and 1.5 percentage points, respectively. Particularly for Antigua and Barbuda and Jamaica, this is certainly not sufficient breathing room to make their governments feel comfortable. Barbados is the only country for which the primary balance to GDP ratio in the baseline scenario (2.7 percent) is significantly larger than that of the default scenario (–0.9 percent).

Target Scenario

The target scenario indicates the primary balance that each country will have to achieve if a sustainable debt/GDP ratio of 60 percent is to be

Table 6.4 Default scenario

Countries	Debt-to-GDP Trajectory									Default Threshold	Baseline Assumptions*			
	2012	2013	2014	2015	2016	2017	2018	2019	2020	≥ Debt Ratio	Exchange Rate**	Interest Rate	Real GDP	Primary Balance
Antigua and Barbuda	76.9	75.4	73.0	71.1	68.6	65.5	61.5	57.2	53.1	77.0	0.0	3.8	3.8	0.8
Belize	81.5	84.5	87.0	88.4	89.6	90.1	90.0	90.3	91.0	91.0	0.0	6.5	2.4	0.4
Barbados	101.0	104.6	105.6	108.7	111.4	114.6	118.8	121.5	122.1	122.0	0.0	6.7	2.6	−0.9
Grenada	87.0	91.2	94.6	98.9	101.0	102.9	105.8	108.7	111.9	76.6	0.0	4.6	2.5	−3.4
St Kitts and Nevis	114.4	113.8	114.7	114.9	113.3	109.4	104.4	98.1	91.3	113.0	0.0	2.7	2.1	3.2
Jamaica	128.2	127.5	127.8	127.2	127.9	129.5	131.6	131.8	130.1	130.0	5.0	8.0	1.4	2.6
Dominica	75.8	81.5	82.9	82.9	82.7	81.0	78.5	75.6	72.2	82.0	0.0	2.8	1.4	−0.9

Notes: *In absolute percentages except for the primary balance which is in percent of GDP.
**Percent depreciation for each country's currency.

Source(s): Authors' estimates.

Table 6.5 Target scenario (debt-to-GDP ratio of 60%)

Countries	Debt-to-GDP Trajectory									Default Threshold	Baseline Assumptions*			
	2012	2013	2014	2015	2016	2017	2018	2019	2020	≥ Debt Ratio	Exchange Rate**	Interest Rate	Real GDP	Primary Balance
Antigua and Barbuda	76.9	75.4	73.0	71.1	68.6	65.5	61.5	57.2	53.1	77.0	0.0	3.8	3.8	0.8
Belize	78.2	77.8	76.6	74.3	71.7	68.6	65.0	61.7	58.8	91.0	0.0	6.5	2.4	3.3
Barbados	94.3	91.3	86.0	82.4	78.1	73.9	70.0	65.1	58.8	122.0	0.0	6.7	2.6	5.2
Grenada	80.9	79.2	76.8	75.1	71.6	68.0	65.0	62.1	59.5	76.6	0.0	4.6	2.5	2.5
St Kitts and Nevis	85.5	80.9	78.0	74.4	69.7	64.0	58.0	51.4	45.0	113.0	0.0	2.7	2.1	3.4
Jamaica	120.5	112.1	104.4	95.8	88.3	81.2	74.1	65.6	55.6	130.0	5.0	8.0	1.4	11.1
Dominica	82.0	84.2	82.5	79.7	77.0	73.2	68.8	64.3	59.7	82.0	0.0	2.8	1.4	2.9

Notes: *In absolute percentages except for the primary balance which is in percent of GDP.
**Percent depreciation for each country's currency.

Source: Authors' estimates.

realized by 2020. The baseline forecasts for the macroeconomic variables are again reused. The results are presented in table 6.5.

To meet the targeted debt/GDP ratio by 2020, Jamaica has to make the largest adjustment to the primary balance. Jamaica has to run a primary surplus of approximately 11.1 percent of GDP, which is considerably larger than the 4.1 percent projected in the baseline scenario. Barbados also requires a relatively large primary balance of 5.2 percent of GDP to meet the target, but it must be noted that in the baseline scenario her primary balance was already projected to be 2.7 percent of GDP. By contrast, whereas Grenada only requires a primary balance of 2.5 percent of GDP to meet the sustainability target, that country was running a primary deficit of 3.4 percent in the baseline scenario, and so requires significant adjustment. Dominica is similarly required to move from a deficit of 0.9 percent in the baseline scenario, to a primary surplus of 2.9 percent of GDP to achieve the targeted debt/GDP ratio. The other countries in the study require relatively small adjustments to their primary balances to achieve the 60 percent debt/GDP target, with St. Kitts and Nevis projected to meet the target with the same primary balance that was incorporated into the baseline.

Conclusion

This chapter has shown that all the heavily indebted Caribbean countries studied, except for Grenada and Dominica, have implemented policies that significantly reduce the risk of default. Grenada and Dominica need to implement additional fiscal austerity measures to strengthen the primary balance so as to enhance the likelihood of solvency. Although Jamaica and Belize have projected debt ratios that fall below the default threshold, a high risk of default exists in these countries because of high refinancing costs in the case of Belize, and a high volume of maturing debt in the case of Jamaica. Fiscal policy in both Jamaica and Belize should therefore be focused on creating more fiscal room to prevent a liquidity crunch.

Notwithstanding the positive results for the other countries, there is limited fiscal flexibility in all the countries examined. A reversal of the fiscal policies implemented since 2008 leads to a breach of the default threshold for all the cases except Belize. However, even in Belize, caution and political strength is called for. High refinance risks could precipitate a liquidity crunch that would presage a default. Most countries have very limited flexibility around policy implementation, with the difference between the projected baseline and default primary balances being very small. Barbados has the greatest flexibility among the countries studied, due to its relatively high debt-carrying capacity, given the depth of the domestic capital market and the greater degree of market confidence.

If the countries investigated in this chapter are to achieve the targeted debt/GDP ratio of 60 percent, the most significant policy adjustments would have to be made by Jamaica, Grenada, and Dominica. While smaller, the requisite adjustments to the primary balance in Antigua and Barbuda and Barbados are not negligible and should be carefully planned. Belize requires a fairly small adjustment, but it faces the aforementioned high refinance risks. Even St. Kitts and Nevis, which does not have to make any adjustments in the projection period to meet the targeted debt/GDP ratio, has to ensure that all the policy measures included in the baseline scenario are implemented, as it was shown to be susceptible to significantly higher than projected debt/GDP ratios if the planned debt-for-land swap is not successful.

So what are the more general lessons for SIDS? The most obvious is that for highly indebted countries, doing nothing about debt is not a viable option. Countries may be tempted to roll-back or delay implementation of stringent fiscal policies due to social pressures and/or political exigencies, but this is highly likely to cause a default on debt obligations. Even in countries where debt restructuring has led to reduced debt levels, projections suggest that fiscal profligacy will precipitate rapid debt accumulation and a speedy return to unsustainability.

Fiscal austerity, however, cannot be the only answer for highly indebted SIDS. This is illustrated for the SIDS examined by the large difference between the baseline forecast for fiscal balances and the targeted balance if sustainable levels of debt are to be achieved. The fact that this large difference persists in spite of austerity measures already being implemented, and does not account for any contingent liabilities that may be incurred, indicates that the fiscal surpluses required to achieve a debt to GDP ratio of 60 percent for highly indebted SIDS may be socially and politically infeasible in the absence of additional measures. Also instructive is the fact that the countries that were closest to the targeted fiscal balance had implemented debt restructuring initiatives at the start of the projection period. The experiences of SIDS with successful debt restructuring initiatives should thus be closely examined for best practices that will minimize the adverse impacts of defaults, while creating much needed breathing space for countries with few, if any, other options.

Finally, the link between debt sustainability and the challenges that SIDS face in achieving sustainable development has also been highlighted in the projections. In spite of the implementation of fiscal austerity measures and the use of debt restructuring, fiscal and debt sustainability in many SIDS is highly vulnerable to economic shocks. Economic growth is stubbornly limited in some SIDS and is quite volatile in most. The openness of many SIDS, along with high reliance on foreign currency denominated debt and increasing reliance on commercial debt also exposes them to exchange rate and interest rate shocks. The sensitivity analyses showed significantly worsened fiscal and debt outcomes for most countries when even moderate shocks were introduced, with the vulnerability to a particular type of shock varying based on the composition of the debt portfolio and nature of the economy. SIDS' vulnerability to natural disasters and climate change exigencies only exacerbate their precarious fiscal and debt positions.

The projections conducted in this chapter suggest that the fiscal austerity measures currently being implemented in many highly indebted SIDS, along with the ad-hoc debt restructuring initiatives, are likely, *ceteris paribus*, to reduce the levels of indebtedness in many of the countries, albeit not to the targeted levels. For SIDS, however, the *ceteris paribus* assumption is always highly likely to be violated because of their vulnerability to external shocks. This vulnerability must be assuaged if debt levels in highly indebted SIDS are to be sustainable in the foreseeable future.

Notes

*The material in this chapter largely reproduces one by the same authors and published as "Chapter 5: Prospects—Extrapolating the Future" in Caribbean Development Bank (2013).

1. Hurley (2011, 1).
2. Hurley (2011, 1).
3. www.imf.org/external/np/exr/facts/jdsf.htm.
4. In spite of challenges to their method that highlighted errors, this result is confirmed in a subsequent errata (Reinhart and Rogoff 2013). http://www.carmenreinhart.com/user_uploads/data/36_data.pdf.
5. IMF (2002).
6. IMF (2011d).
7. IMF (2011d).
8. IMF (2002, 3).
9. See Geither (2003) and Allen (2005) for modifications to this approach.
10. Appendix appears at the end of chapter 7.
11. The estimate of market support refers to the amount of money a government can borrow from capital markets.
12. The data used in these analyses were collated from various country reports, statistical digests, budget statements, IMF reports, and regional reports.
13. See MTFS 2010–2014, pages 11–16.
14. Although there was a 96.8 percent voluntary participation rate, the activation of the collective action clauses (CACs) embedded in exchanged instruments leads to 100 percent participation. Fiscal authorities have expressed their intent to exercise the CAC clauses.

References

Allen, M. 2005. *Information Note on Modifications to the Fund's Debt Sustainability Assessment Framework for Market Access Countries.* Washington, DC: International Monetary Fund.

Bhagoo, Vangie and Patrick Watson. 2009. "Predicting a Debt Crisis in Selected Caribbean Countries." Sir Arthur Lewis Institute of Social and Economic Studies, University of the West Indies St. Augustine.

Business Monitor International (BMI). 2012a. *East Caribbean Business Forecast Report.* London: Business Monitor International.

———. 2012b. *West Caribbean Business Forecast Report.* London: Business Monitor International.

Caner, Mehmet, Thomas Grenner, and Fritzi Koehler-Geib. 2010. "Finding the Tipping Point: When Sovereign Debt Turns Bad." *World Bank Policy Research Paper WPS5391.*

Caribbean Development Bank. 2013. *Public Sector Debt in the Caribbean: An Agenda for Reduction and Sustainability.* Bridgetown, Barbados: Caribbean Development Bank.

Central Bank of Barbados. 2011. *Barbados Annual Economic Review 2010.* Bridgetown: Central Bank of Barbados.

———. 2010. *Barbados Economic and Financial Statistics.* Bridgetown: Central Bank of Barbados.

Central Bank of Belize. 2011. *Quarterly Review.* Belize City: Central Bank of Belize. https://www.centralbank.org.bz/publications-research/economic-publications/quarterly-reviews.

Geither, T. 2003. *Sustainability Assessments—Review of Application and Methodological Refinements.* Washington, DC: International Monetary Fund.

Government of Antigua and Barbuda. 2011. *2012 Budget Statement.* http://www.antigua.gov.ag/pdf/finance/buget_statement_2012.pdf.

Hurley, Gail. 2011. "Addressing Unsustainable Debt in Small Island Developing States." *One Pager No. 123*, International Policy Center for Inclusive Growth.

International Monetary Fund (IMF). 2002. *Assessing Sustainability.* Washington, DC: International Monetary Fund. https://www.imf.org/external/np/pdr/sus/2002/eng/052802.pdf.

———. 2011a. *Barbados: 2011 IMF Article IV Consultation, IMF Country Report No. 12/7.* Washington, DC: International Monetary Fund. http://www.imf.org/external/pubs/ft/scr/2012/cr1207.pdf.

———. 2011b. *Belize: 2011 Article IV Consultation, IMF Country Report No. 11/340.* Washington, DC: International Monetary Fund. http://www.imf.org/external/pubs/ft/scr/2011/cr11340.pdf.

———. 2011c. *Dominica: 2011 Article IV Consultation, IMF Country Report No. 11/324.* Washington, DC: International Monetary Fund. http://www.imf.org/external/pubs/ft/scr/2011/cr11324.pdf.

———. 2011d. *Modernizing the Framework for Fiscal Policy and Public Debt Sustainability Analysis.* Washington, DC: International Monetary Fund. http://www.imf.org/external/np/pp/eng/2011/080511.pdf.

Pescatori, Andrea, Damiano Sandri, and John Simon. 2014. "Debt and Growth: Is There a Magic Threshold?" *IMF Working Paper WP/14/34.* Washington, DC: International Monetary Fund. http://www.imf.org/external/pubs/ft/wp/2014/wp1434.pdf.

Reinhart, Carmen M and Kenneth S. Rogof. 2010. "Growth in a Time of Debt." *NBER Working Paper Series 15639.*

Sahay, R. 2005. "Stabilization, Debt, and Fiscal Policy in the Caribbean." *IMF Working Paper WP/05/26.*Washington, DC: International Monetary Fund. http://www.michelerobinson.net/yahoo_site_admin/assets/docs/IMF_Stabilization_Debt_and_Fiscal_Policy.13654209.pdf.

Chapter 7

Causes of Debt Accumulation in SIDS

Damien King

This chapter investigates the drivers of debt accumulation in Small Island Developing States (SIDS). The question of what drives small economies to accumulate debt on the scale observed in the last quarter century remains an open one in the face of both commonalities and differences in the circumstances and characteristics of these economies. The islands in the Caribbean share a relatively small geographical space—the latitudes of the Caribbean islands span only some 2,200 kilometers—and so are exposed to common climatic and other natural conditions. The geographical proximity also explains the similarities in the economic structures. Most of the islands are dependent on tourism and primary commodities. In contrast the islands of the AIMS region are dispersed across a latitudinal span, from Cabo Verde to Singapore, of roughly 14,000 kilometers. Their natural and economic conditions vary widely.

At the same time, SIDS, by definition, share the characteristics of being small and being islands, characteristics that force upon them a narrow economic structure and the choice between high public administration costs or weak administrative capacity, which trade-off results from their inability to exploit economics of scale in public administration. It remains to be established, therefore, whether their commonalities drive their debt accumulation or if their different circumstances defy the finding of a common explanation.

There are many candidate explanations for the tendency to indebtedness of small islands and all have been advanced in the literature at some point: external economic shocks, natural disasters, limited natural resources, distance from markets, fiscal indiscipline, contingent liabilities, slow growth, and debt dynamics. This chapter seeks to sort through that thicket by identifying the proximate cause of debt accumulations in SIDS using a commonly deployed decomposition methodology and applying it to the particular small islands and periods over the last 25 years in which debt has risen significantly over short periods of time.

That analysis reveals whether the debt originates within or outside of central government and if the debt was used to finance development or to plug gaps created by fiscal slippage.

The analysis herein includes both external and domestic debt. The increased depth of domestic capital markets, even in small island economies, has allowed local savers to absorb increasing amounts of their government's paper, so the domestic debt is a significant portion of the total for some islands. In addition, Panizza (2008) has argued that the focus on external debt is increasingly misplaced since the integration of capital markets renders the sharp distinction between the two types of debt increasingly redundant.

The rest of the chapter is organized as follows. The first section provides some background on the state of the literature as well as on the methodology. The next two provide the results of the decomposition, first for the Caribbean and then for the AIMS and Pacific regions. In the final section, we draw some tentative conclusions based on the observed patterns in the analysis.

Context and Methodology

While the debt problem of small states often arises in the literature, it is in the context of broad discussions of their economic challenges (Briguglio et al. 2006; Favaro et al. 2011). There has been very little work in the way of an investigation of the drivers of debt in SIDS in general and within that, the Caribbean, the most indebted, has received nearly all of the attention (Perez 2007; Sahay 2005).

To explain the high level of indebtedness in the Caribbean, Perez uses a flow-of-funds model that aggregates the fiscal account and thus is not designed to ferret out drivers on the fiscal side. Sahay finds that primary deficits are the major cause with significant supporting roles for interest payments and the stock-flow adjustment component. Dodhia (2008) attempts a comprehensive analysis of the problem on a global scale, but in discussing causes he does not employ a decomposition and so is unable to rigorously determine the sources of indebtedness. Instead, he examines contemporaneous macroeconomic developments and tries gamely to establish links between those and rising debt. But this approach is unable to establish the relative contribution of the different factors and so ends up with a long list of causes, nonetheless identifying primary deficits as the main one.

The International Monetary Fund (2013) investigates the debt problem on the global scale of Dodhia (and the present paper) and with a decomposition similar to the one employed here, but was applied only to the years from 2007 to 2011. Despite sub-dividing their sample into small and micro-states, the authors find that debt accumulation in both sets of countries was driven by primary deficits and interest payments.

This paper adds to the literature on debt in SIDS by employing the commonly used accounting decomposition to small island economies across the globe in order to derive comprehensive conclusions on debt accumulation while teasing out regional differences. The analysis is applied to 13 SIDS drawn from all three SIDS regions around the world. The selected countries are those that have experienced sustained increases in debt, that is, lasting at least three consecutive years, at any point over the last quarter century and for which the requisite data could be obtained. The criterion yields the following: from the Caribbean, Antigua and Barbuda, Barbados, Dominica, Grenada, Jamaica, and St. Kitts and Nevis; from AIMS: Cabo Verde, Comoros Islands, Cyprus, Malta, and the Seychelles; and from the Pacific: Samoa and the Solomon Islands.[1]

The work is based on a debt decomposition method that has become standard in the analysis of debt accumulation (e.g. Budina and Fiess 2005). The decomposition is accomplished by assigning to each component balance of the budget (recurrent, capital, and so on) the amount of financing it would require (or surplus to be disposed of) on an annual basis. The change in debt from one year to the next is thus decomposed into the contributions of (1) the primary fiscal balance—which will require new borrowing to finance the gap between expenditure and revenue; (2) the real interest cost—which will require new borrowing to service the existing debt; (3) revaluations—in which the value of existing debt may change if the unit of measurement differs from the units in which the debt obligation is denominated; and (4) stock-flow adjustments—which usually reflect debt assumed by central government but that was contracted from outside of it, or off-budget liabilities (OBLs). Since the stock-flow adjustment calculation is actually the residual after all identifiable flow variables are measured, it would also sweep up any accounting/measurement errors and for that reason merits some degree of caution in interpretation.

The primary balance component in the tables aggregates the recurrent elements of the fiscal budget with the capital portion. The former includes tax and other recurring revenue along with non-capital expenditure while the capital portion of the budget is where one finds grants and asset sales on the revenue side and capital expenditure and asset purchases on the spending side. Revaluations in the stock of existing debt can come from two sources. Changes in the official nominal exchange rate will inflate or deflate the domestic currency representation of the external portion of the debt. In addition, since the debt is being measured in "GDP units," then a rise or fall in GDP will also deflate or inflate the debt/GDP ratio accordingly.

The decomposition application below employs a methodological innovation. The interest contribution is usually calculated in real terms conventionally defined as the difference between the implicit nominal interest rate on debt and the rate of inflation usually measured by the GDP deflator. The result of that adjustment is designated as real because any change in the value of the unit of account is taken into account—the unit of account being the local currency. However, in an analysis of changes in the debt to GDP ratio, the unit of account for the debt is nominal GDP units. The nominal GDP changes because of price inflation but also because of real economic growth. Thus, if the nominal rate of interest is to be excised of the effect of changes in the unit of account, then both the GDP deflator and the real rate of economic growth ought to be deducted.

Another way to understand the definition of the real interest rate used here is to recognize that if there are no other contributors to public debt except for new borrowing to pay interest, then as long as the average nominal interest rate is equal to the growth rate of nominal GDP, the debt to GDP ratio will not change. Therefore, the real rate of interest would be zero. Accordingly, the real rate of interest applicable to this analysis is the nominal rate net of changes in the nominal GDP, which means the portion greater than both inflation and real GDP growth. That is the definition of the real rate of interest used in the decompositions presented below.

In order to draw lessons for what underlies debt accumulation in SIDS, the present analysis diverges from the existing literature by the use of non-contemporaneous time periods. Previous cross-country studies select a particular time span and examine the changes in debt during that same time period for all the countries in the analysis (Bandiera 2008; International Monetary Fund 2013; Sahay 2005). The objective here is to understand why debt tends to rise whenever that may be, rather than whatever happened to occur during any particular time period. Accordingly, the decompositions are applied to each of the countries only for the time periods during which debt was rising in that country. This focuses the analytical spotlight on the periods when the problem occurs and does not pollute the analysis with periods when debt is not rising.

This focus on each country's experience only when its debt is rising is important since the periods of debt accumulation for the various states do not generally overlap. Figure 7.1 displays the years of significant debt accumulation for each of the dozen countries in the analysis. While there is a bunching of adverse debt movements in the 1996 to 2004 period, there are notable outlying movements that are out of synchronization with that period. Both Cabo Verde and the Seychelles had significant debt accumulation in the early nineties instead of during the

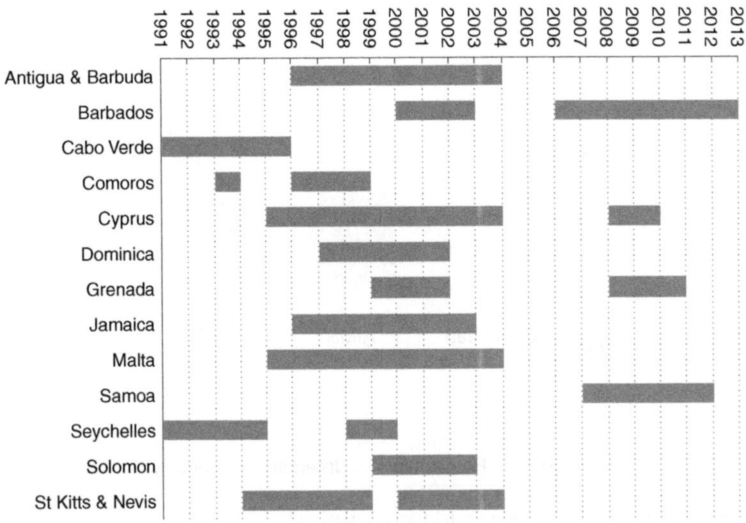

Figure 7.1 Periods of rising debt in SIDS, 1991–2013

modal period around the turn of the century. And Barbados, Grenada, and Cyprus had notable debt increases coming on to the end of the last decade and the beginning of the present one.

General Observations

Summing the annual decompositions for the selected years for each country yields the decomposition by country. Then averaging across the countries in each of our three regions provides the regional results, presented in figure 7.2. It shows that the primary driver of debt in small islands differs markedly by region. In the six Caribbean islands in this study, the main factor, overwhelmingly, appears to be off-budget liabilities—debts assumed from the contingent and contracted obligations of public sector enterprises, public–private partnerships, and in some cases, the private sector. OBLs accounted for 53 GDP percentage points of debt while the second most important factor, interest payments, was responsible for only 18 GDP percentage points. Put another way, OBLs explain three-fifths of the debt increases in the Caribbean.

The other two regions show a pattern distinct from the Caribbean but similar to each other. The main driver of debt accumulation in both AIMS (five islands) and the Pacific (two islands) is primary deficits, accounting for 31 and 48 percentage points of debt respectively. No other factor is close in quantitative significance while the role of OBLs is negligible.

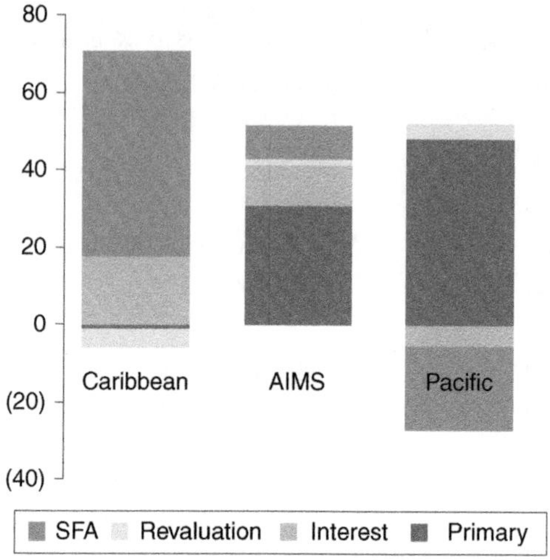

Figure 7.2 Decomposition of debt accumulation by region

While the purpose of this analysis is not to measure absolute debt levels across the three regions, it is noteworthy that merely selecting at least three consecutive years of debt growth in islands in the three regions reproduces the relative quantities of debt that corresponds to the comparative severity of the debt problem in the regions. The decompositions account for an average of 59 percentage points of debt in the Caribbean, 54 in AIMS, and 31 in the Pacific.

Caribbean

The remarkable revelation of the exercise is that OBLs were the major factor in every instance of significant debt increase in every Caribbean island in the sample except for Barbados (and even there it was the second largest contributor behind interest payments—table 7.1). The cost of these liabilities that central governments take over ranged from Barbados' low of 24 GDP percentage points to 84 and 85 for St. Kitts and Nevis and Jamaica, respectively. This is clearly a systemic problem in the Caribbean and speaks to the need for improved monitoring and management of explicit guarantees and other contingent liabilities, particularly in relation to public enterprises.

Within the Caribbean context, the sources of these OBLs are mainly public enterprises and private financial sector failures, the latter being a significant contributor. In Jamaica, the financial crisis of the late 1990s contributed 35 GDP percentage points of debt; the collapse of

Table 7.1 Debt decomposition for the Caribbean region

	Primary	Interest	Revaluation	OBL
Antigua & Barbuda	12	14	(19)	51
Barbados	(16)	36	5	24
Dominica	8	6	(3)	34
Grenada	20	4	7	41
Jamaica	(55)	38	(2)	84
St Kitts & Nevis	27	8	(20)	85
Average	**(1)**	**18**	**(5)**	**53**

Note: Units = percentage points of GDP.

Source(s): Author's calculations.

the Colonial Life Insurance Company in Trinidad and Tobago in the late 2000s resulted in losses that had to be absorbed by nearly every government in the English-speaking Caribbean. Public enterprise losses absorbed by central governments have included sugar companies in St. Kitts/Nevis and Jamaica, the national airline in Jamaica, and a sports development company and public–private partnerships in Barbados.

Far behind the overwhelming contribution to debt from OBLs was the contribution of interest payments on existing debt. The cost of servicing debt follows from merely having debt, and since the Caribbean is the most indebted of the small island regions, it naturally follows that the region's interest cost would be a larger contributor than for the less indebted regions of small islands.

Beyond the minimum necessary cost of servicing debt, though, there is evidence that the indebted countries of the region did not all manage their debt equally well in relation to cost minimization. The range of implicit average interest rates on their debt extends from a low of 3 percent in Grenada to a high of 12 percent in Jamaica, even though many of these countries were indebted at the same time in the same international capital environment. It suggests that at least some part of the high contribution of interest cost to sustaining high debt levels was due to debt management choices. This applies particularly to Antigua and Barbuda and Jamaica which seem to have made more costly choices than other islands. Interest payments contributed 38 percentage points to Jamaica's debt, more than any other island anywhere in the world. In the Caribbean, it also explained 36 percentage points of Barbados' debt and 14 percentage points of Antigua and Barbuda's. For the remaining islands, interest costs were not significant.

Primary deficits are not revealed to be a major problem in the Caribbean at the aggregate level, but the aggregation masks wide differences. Table 7.1 shows that Jamaica and Barbados were accumulating debt despite running large primary surpluses. At the other end, primary deficits in Grenada and St. Kitts and Nevis accounted for at least a quarter of their total debt accumulation.

Upward revaluation of the existing debt stock, either due to exchange rate depreciation or GDP contraction, was not a contributor to indebtedness in the Caribbean. Antigua and Barbuda, Dominica, Grenada, and St. Kitts and Nevis are members of the Eastern Caribbean Currency Union and the shared currency maintains a stable peg to the US dollar, underpinning a generally stable, low inflation environment. Though not in the currency union, both the stable peg and low inflation are also true for Barbados. There was little room, therefore, for real currency movements to contribute materially to debt revaluations in these islands.

AIMS and Pacific Islands

The decompositions reveal that the indebted islands of both the AIMS and Pacific regions became so mainly because of fiscal slippage. Table 7.2 reveals that primary deficits were the largest contributor to debt in five of the seven islands in the sample drawn from these two regions (Cabo Verde, Cyprus, Malta, Samoa, and the Solomon Islands). In these two groups, no other source was significant. Liabilities originating outside of the budget, which played a key role in debt accumulation in the Caribbean, in general play no significant role in either region. Interest payments explains a fifth of the debt in AIMS but for the Pacific islands makes a negative contribution in real terms. It is not surprising that the AIMS and Pacific countries, with lower debt levels than in the

Table 7.2 Debt decomposition for the AIMS and Pacific regions

	Primary	Interest	Revaluation	OBL
AIMS				
Cabo Verde	62	(6)	(16)	6
Comoros	8	(4)	30	4
Cyprus	50	15	(3)	(32)
Malta	35	17	(11)	(4)
Seychelles	(1)	29	8	69
Average	**31**	**10**	**2**	**9**
PACIFIC				
Samoa	56	(3)	(9)	(22)
Solomon Islands	40	(8)	18	(21)
Average	**36**	**6**	**2**	**0**

Note: Units = percentage points of GDP.

Source(s): Author's calculations.

Caribbean, do not show a significant contribution of interest payments to debt accumulation.

Revaluations play a positive but small role in both regions. Given that the islands in the AIMS and Pacific regions are more dependent on external debt than those in the Caribbean, it seems surprising that the greater exchange rate exposure seems not to have resulted in a noticeable contribution of revaluations to debt growth. Recall, though, that debt can be revalued in two ways—exchange rate changes and GDP fluctuations. A disaggregation of those two channels does reveal that while exchange rate depreciations have indeed pushed debt upward, by six percentage points on average, economic growth has pushed it down at the same time by nearly as much. As table 7.3 shows, however, which of the two forces dominates varies greatly from island to island.

The contrast with the findings for the Caribbean is even more striking than the averages suggest because the Seychelles is an outlier in AIMS. Its debt decomposition more closely resembles those in the Caribbean with OBLs and interest cost accounting for nearly all of the debt during the six years of significant debt growth included in our data set. Leaving the results from the Seychelles out of the average, primary deficits account

Table 7.3 Disaggregation of contributions to revaluations

	Total Revaluations	Exchange Rate Depreciation	Economic Growth
Caribbean			
Antigua & Barbuda	(19)	(24)	5
Barbados	5	0	5
Dominica	(3)	(4)	1
Grenada	7	2	5
Jamaica	(2)	(6)	4
St Kitts & Nevis	(20)	(18)	(2)
Average		**(8)**	**3**
AIMS			
Cabo Verde	(16)	(12)	(4)
Comoros	30	(3)	33
Cyprus	(3)	3	(5)
Malta	(11)	(11)	0
Seychelles	8	(20)	28
Average		**(9)**	**10**
Pacific			
Samoa	(9)	(2)	(7)
Solomon Islands	18	8	9
Average		**3**	**1**

Note: Units = percentage points of GDP.

Source(s): Author's calculations.

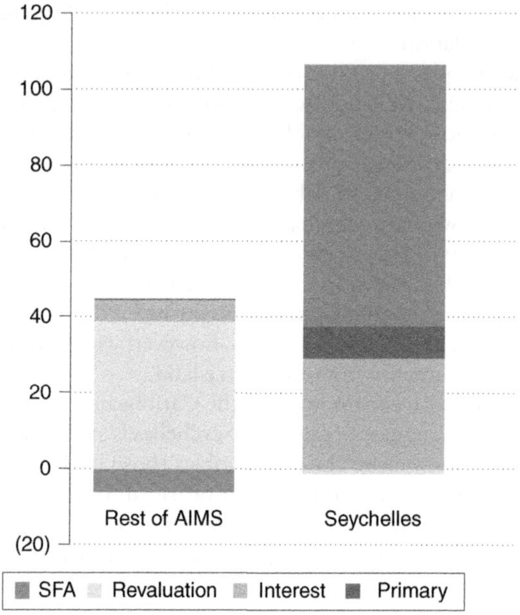

Figure 7.3 Decomposition of debt accumulation, Seychelles and the rest of AIMS

for nearly all of the debt increases in the remaining islands in our sample from AIMS. Figure 7.3 separates the results for the Seychelles and the remainder of AIMS islands. Moreover, without the Seychelles in the AIMS results, the OBL term, which, recall, is a residual that reconciles stock-flow discrepancies, becomes negative, meaning it reduces debt. Note also that the OBL term is also negative for the Pacific islands. This is because some capital expenditure is grant-financed and not all of it may flow through the revenue side of the budget, especially in Samoa.

In the context of the AIMS and Pacific regions, we have noted that the Seychelles is an outlier. Our data identify two periods of debt growth in the Seychelles, from 1991 and 1995 and again between 1997 and 1999. The events that make this island an anomaly in the context of its region are all derived from the first period, in which three-quarters of its 65 percentage points of GDP debt increase originated from outside the central government. The second episode conforms to the pattern of the other islands in the region.

The difference in the patterns of debt accumulation among the regions, along with the near consistency of the results within each of them, begs for an explanation. The Caribbean's higher exposure to and realization of contingent liabilities raises questions about the institutional character and nature of governing relative to the situation that surrounds the islands in AIMS and the Pacific. That is the subject of other work (Robinson 2014).

Capital Expenditure

Analysis of the decompositions at a more disaggregated level highlights the role that public capital expenditure plays in debt accumulation. The exercise reveals that it plays a different role in each region, quantitatively and qualitatively.

Borrowing to finance primary deficits, as is the case in AIMS and the Pacific region, can be a greater or lesser concern depending on the nature of the expenditure that was financed. A further disaggregation of the role of primary deficits reveals that the capital portion of the budget accounts for nearly all of the deficits (table 7.4). With the exception of the Comoros Islands and the Solomon Islands, both of which had significant primary recurrent deficits during their periods of debt accumulation, all the remaining countries—Cabo Verde, Cyprus, Malta, the Seychelles, and Samoa—were mostly borrowing to invest.

Table 7.4 Disaggregation of contributions to primary balances

	Total Primary Balance	Primary Recurrent Balance	Capital Balance
Caribbean			
Antigua & Barbuda	12	0	12
Barbados	(16)	(43)	27
Dominica	8	(9)	16
Grenada	20	(21)	41
Jamaica	(55)	(67)	12
St Kitts & Nevis	27	(26)	53
Average		**(28)**	27
ROW			
Cabo Verde	62	(17)	79
Comoros	8	17	(9)
Cyprus	50	8	43
Malta	35	0	34
Seychelles	(1)	(41)	40
Average		(7)	37
Pacific			
Samoa	56	4	52
Solomon Islands	40	24	16
Average		**14**	**34**

Note: Units = percentage points of GDP.

Source(s): Author's calculations.

This finding, that small islands outside the Caribbean borrowed to invest while those in the Caribbean realized contingent liabilities, is at least consistent with the observation that the non-Caribbean groupings achieved faster growth rates than the Caribbean six. For the decade of the 2000s, while the annual growth rates of the indebted Caribbean economies averaged 1 percent and occupied the range from zero (Barbados) to 2.4 percent (Grenada), the average for the non-Caribbean countries was 3.4 percent, ranging from 1.5 percent (Malta) to 5.9 percent (Cabo Verde).

The decompositions revealed that, in the Caribbean, in only three islands (Antigua and Barbuda, Grenada, and St. Kitts and Nevis) did primary deficits contribute to debt buildup. In all of them, as was the case in the AIMS and Pacific countries, it was not an imbalance on the "housekeeping" recurrent side of the budget that created the deficit but rather on the capital side. But there is an important difference. The sporadic nature of the public capital expenditure in the Caribbean hints at its cause—reconstructive capital expenditure following natural disasters.

Hurricane Jose that struck St. Kitts and Nevis in 1999 necessitated a 50 percent increase in public capital expenditure over the next two years. The doubling of Grenada's debt/GDP over the first three years of the new millennium, from 54 percent to 110, followed by an 80 percent rise in public capital expenditure during that period, was necessitated by reconstruction following two devastating hurricanes hitting the island in 2000 and 2002. In 2008, Jamaica's public capital expenditure had to increase by 76 percent because of Hurricane Dean.

The examples expose the importance of vulnerability in getting a full understanding of the debt problem in small islands. In many of the instances, then, in which Caribbean islands run primary deficits that contribute to debt build-up, it is not due to discretionary fiscal slippage but rather to another kind of contingent liability, that derived from the region's vulnerability to natural disasters. This means that the role of contingent risks in debt accumulation in the Caribbean is even greater than the amount reflected in the estimated size of the considerable OBL contribution.

Conclusion

A striking result of this analysis is that the drivers of debt accumulation in small islands show clear and distinct regional patterns. The damaging role of contingent risks in the Caribbean, both from public and private enterprises and also from the islands' vulnerability to natural disasters, is an important result and a source of deep concern because of its implications for the fiscal governance of the countries in the region. In sharp contrast, for the islands in the rest of the world, the source of debt is to be found mostly within central government operations and seems to be of a development finance nature. This observation at least goes some way toward explaining the vast difference in growth outcomes between the regions.

The finding has two crucial implications for policy. First, if growth is to take place in the Caribbean, then a way of managing the inherent vulnerabilities of the region along with other contingent risks must be found. Greater attention should be paid to insurance schemes for natural disasters and to identifying, estimating, hedging, and managing the risks lurking in public enterprises and public–private partnerships. Since the debt is not itself creating a path to growth that balances the drag of the debt overhang, the region could remain stuck for a long time.

The other implication of the results obtained here is that, for the countries in the Caribbean, the fiscal consolidation that is a standard part of every economic reform package, even if ambitious in design and faithfully implemented, may not be sufficient to ensure debt sustainability. This is because, notwithstanding balanced budgets, these islands will continue to be exposed to all of the contingencies that gave rise to the debt in the first place. So while fiscal slippage is not adding to their debt stock, liabilities outside of the full control of the governments will continue to add new debt obligations even as debt management, restructurings, and primary surpluses keep the old ones in check.

The islands in the AIMS and Pacific regions seem to be in a better position than those in the Caribbean. With thin domestic capital markets, they are unable to generate the surpluses domestically that can support the scale of investment required for economic growth. Some of this capital is provided bilaterally by way of grants, especially in the Pacific Islands. The remainder has given rise to sovereign debt that on occasion gets out of control. More careful scrutiny of development expenditure may lead to more efficient use of capital and thus reduced external borrowing. In any event, greater attention should be paid to rising debt levels.

A more general lesson is to recognize that, while small islands share many characteristics that give rise to common problems, the differences are potentially important too. This paper has uncovered that differences in geography are associated with a clear pattern of differences in fiscal outcomes. It is a lesson that has implications for areas of policy beyond debt management.

Appendix: Debt Decomposition Derivation

Variables

D_t = *total public debt in local currency*

D_t^d = *domestic debt in local currency*

$D_t^f = \textit{external debt in local currency}$

$d_t = \textit{total public debt in GDP units}$

$d_t^f = \textit{external debt in GDP units} = \alpha\, d_t$

$d_t^d = \textit{domestic debt in GDP units} = (1-\alpha)\, d_t$

$e_t = \textit{nominal exchange rate, domestic currency per unit of foreign currency}$

$SFA_t = \textit{new, non-borrowed debt}$

$F_t = \textit{external debt in foreign currency}$

$g_t = \textit{rate of real economic growth}$

$GDP_t = \textit{gross domestic product} = (1+g_t)(1+\pi_t)GDP_{t-1}$

$i_t = \textit{average interest rate on total debt} = (1-\alpha_t)i_t^d + \alpha_t(1+s_t)i_t^f$

$i_t^d = \textit{average interest rate on domestic debt}$

$i_t^f = \textit{average interest rate on external debt}$

$PB_t = \textit{primary balance}$

$s_t = \textit{rate of nominal depreciation, such that: } 1+s_t = \dfrac{e_t}{e_{t-1}}$

$s_t^r = \textit{rate of real depreciation, such that: } 1+s_t^r = (1+s_t)\dfrac{1+\pi}{1+\pi_t}$

$\alpha_t = \textit{share of external in total debt}$

$\pi_t = \textit{GDP deflator}$

Derivation

$$D_t = -PB_t + SFA_t + (1+i_{t-1}^d)D_{t-1}^d + (1+i_{t-1}^f)e_t F_{t-1} \tag{1}$$

Equation (1) states the total debt in any year is the sum of the financing needs of the primary deficit, new non-borrowed debt, and the stock of inherited debt along with interest due on that stock, disaggregated into domestic and external components, all expressed in local currency.

Dividing through by GDP_t and using lower case to represent variables in GDP units yields

$$d_t = -pb_t + sfa_t + \frac{(1+i_{t-1}^d)D_{t-1}^d}{(1+g_t)(1+\pi_t)GDP_{t-1}} + \frac{(1+i_{t-1}^f)e_t F_{t-1}}{(1+g_t)(1+\pi_t)GDP_{t-1}} \tag{2}$$

Substituting the total debt/GDP ratio, d_t, for the domestic and external debt variables, the equation becomes

$$d_t = -pb_t + sfa_t + (1 + i_{t-1} + \alpha_{t-1} s_t)\frac{d_{t-1}}{(1+g_t)(1+\pi_t)} \tag{3}$$

Subtracting d_{t-1} from both sides

$$\Delta d_t = -pb_t + sfa_t + (1 + i_{t-1} + \alpha_{t-1} s_t)\frac{d_{t-1}}{(1+g_t)(1+\pi_t)} - d_{t-1} \tag{4}$$

$$\Delta d_t = -pb_t + sfa_t + [i_{t-1} - \pi_t(1+g_t) - g_t + \alpha_{t-1} s_t]\frac{d_{t-1}}{(1+g_t)(1+\pi_t)} \tag{5}$$

Substituting $s_t = (1+s_t^r)\frac{1+\pi_t}{1+\pi^{\rightarrow}} - 1 = \frac{(\pi_t - \pi^{\rightarrow})}{1+\pi^{\rightarrow}} + s_t^r \frac{1+\pi_t}{1+\pi^{\rightarrow}}$ and rearranging

$$\Delta d_t = -pb_t + sfa_t + \begin{bmatrix} i_{t-1} - \pi_t(1+g_t) + \alpha_{t-1}\frac{(\pi_t - \pi^{\rightarrow})}{1+\pi^{\rightarrow}} \\ -g_t + \alpha_{t-1} s_t^r \frac{1+\pi}{1+\pi^{\rightarrow}} \end{bmatrix} \frac{d_{t-1}}{(1+g_t)(1+\pi_t)} \tag{6}$$

The above equation states that the change in debt is the sum of the following possible sources, in order of appearance: new autonomous borrowing, necessitated by a primary deficit, pb_t; the assumption of existing liabilities external to central government, sfa_t; and debt due to existing debt, which in turn can be due either to the cost of interest on the debt (consisting of the first three terms inside the square brackets) or the revaluation of existing debt stocks. Such a revaluation could come about either because of real economic growth (since we measure the debt in GDP units) or because changes in the real exchange rate revalue the external portion of the debt stock.

The third part of the interest rate effect, the term $\alpha_{t-1}\frac{(\pi_t + \pi^{\rightarrow})}{1+\pi^{\rightarrow}}$, represents the difference between a nominal and real depreciation of the currency.

For the debt decomposition in this chapter, the primary balance, pb_t, is further decomposed into a primary recurrent balance, bb_t, and a capital balance, kb_t.

Note

1. Cyprus is not usually included in a discussion of SIDS and Malta only occasionally, but both are poorer than Singapore, which is always included. And their debt experiences have at least cursory similarity to other small island states, so they have been included.

References

Bandiera, Luca. 2008. "Public Debt and Its Determinants in Low Income Countries—Results from 7 Country Case Studies." World Bank Working Paper Series.

Briguglio, Lino, Gordon Cordina, and Eliawony J. Kisanga. 2006. *Building the Economic Resilience of Small States.* Malta: Formatek Ltd.

Budina, Nina and Norbert Fiess. 2005. *Public Debt and Its Determinants in Market Access Countries: Results from 15 Country Case Studies.* Unpublished paper. Washington, DC: World Bank.

Dodhia, Dinesh. 2008. "The Emerging Debt Problems of Small States." Economic Paper Series. London: Commonwealth Secretariat.

Favaro, Edgardo, Dörte Dömeland, William O'Boyle, and Tihomir Stucka. 2011. "Small States, the Financial Crisis, and the Aftermath." In *Sovereign Debt and the Financial Crisis: Will This Time Be Different*, edited by Carlos A. Primo Braga and Gallina A. Vincelette. Washington, DC. The World Bank.

International Monetary Fund (IMF). 2013. *Macroeconomics Issues in Small States and Implications for Fund Engagement.* Washington, DC: International Monetary Fund.

Panizza, Ugo. 2008. "Domestic and External Public Debt in Developing Countries." *UNCTAD Discussion Papers 188.* Geneva: United Nations Conference on Trade and Development.

Perez, Estaban. 2007. *Debt Accumulation in the Caribbean.* New York: United Nations.

Robinson, Michele. 2014. "Institutional Underpinnings of Debt in SIDS." In *Debt and Development in Small Island Developing States*, edited by Damien King and David Tennant. Basingstoke: Palgrave McMillan.

Sahay, Ratna. 2005. *Stabilization, Debt, and Fiscal Policy in the Caribbean.* Washington, DC: International Monetary Fund.

Chapter 8

Institutional Underpinnings of Debt in SIDS

Michele Robinson

Previous chapters have shown that many Small Island Developing States (SIDS) are burdened with unsustainable levels of debt with widening fiscal deficits and contingent liabilities accounting for much of the debt accumulation. This raises questions about governance and the institutional framework for public debt management in SIDS, and whether weak institutional structures for public debt management play a role in public debt accumulation and in creating risky debt structures.

This chapter explores this issue. It examines how the political structure heightens the likelihood of widening deficits and high levels of contingent liabilities in SIDS. It then moves on to discuss the institutional framework for public debt management in the context of best practice guidelines developed by the leading international financial institutions in conjunction with debt management experts and practitioners worldwide. It then relates these guidelines to the existing institutional structures in SIDS. The chapter focuses, in particular, on the legal and organizational framework for public debt management, highlighting the challenges to debt management imposed by weak institutional structures and examining how they may have contributed to the high debt burdens that have plagued SIDS for more than a decade. The chapter concludes by suggesting ways in which SIDS, faced with resource constraints and limited capacity, can strengthen their institutional framework to ensure more effective public debt management and a sustainable debt trajectory.

The Impact of Political Structures on Debt Accumulation and Public Debt Management

While the previous chapters have shown the effect of large and growing fiscal deficits on debt accumulation, less has been discussed about the political and institutional structures underlying weak fiscal policy and debt management. This chapter finds that the quality of public institutions plays a significant role in public debt accumulation. In SIDS where institutional structures are weak and government is dominant, excessive levels of debt are more likely to occur.

The Role and Size of Government in SIDS

Brown (2009) argues that, contrary to a school of thought that posits that small size facilitates good governance and provides an environment conducive to successful institutional development, the evidence shows that small states are characterized by organizational structures that are "hierarchical, bureaucratic, rigid and inflexible" and "leadership—at both political and bureaucratic levels—that is paternalistic at best, autocratic at worst" (Brown 2009, 8). He finds further that "personal considerations, including nepotism, are the prevalent, if not primary, basis for personnel transactions rather than merit or performance, potentially leading to corruption" (Brown 2009, 8). These characteristics undermine good governance and public administration in small states and translate into weak public policy. It is weak fiscal policy in particular, and the difficulties in controlling the size of fiscal deficits that have raised public debt to unsustainable levels and placed a majority of SIDS, particularly those in the Caribbean, at risk of high debt distress.

A common characteristic of small states is the dominance of the public sector both in terms of expenditure and employment. Several studies have shown that the size of government, as measured by the government expenditure to GDP ratio, is larger in small states than in large states (Amo-Yartey et al. 2012; International Monetary Fund 2013). As has been argued by Brown (2009), from "cradle to grave" citizens rely heavily on the state for the provision of social services and employment because of weak private sectors and weak civil society institutions. Thus the state has high expenditures, not only in terms of the provision of public goods, but in terms of its wage bill as the government is typically the largest employer in the state. Total government expenditures in small states averaged close to 40 percent of GDP in the mid-2000s (Medina Cas and Rui 2008), while for the Caribbean total expenditures averaged around 30 percent of GDP with the wage bill accounting for the most significant share.

Governments in small states tend therefore to assume a highly paternalistic role given the relatively large size of the public sector and the population's dependence on a wide array of public goods and services, including social housing, health and education, social protection for vulnerable groups and employment. In such an environment, where political divisions are often deep and differences are "more personal, more intense and more emotionally charged" (Brown 2009, 9), public policy decisions are therefore often less guided by the probity of economic decisions than by the need to meet the demands of a vocal citizenry who can readily excise governments by exercising their power at the polls. As a result, public policy often subordinates fiscal prudence to political expediency.

The subordination of sound public policy in favor of political priorities is manifest in the high wage bill and government's role as employer of last resort; the high share of contingent liabilities and government's dominant role in promoting economic growth and development; government's resistance to those institutional and legislative reforms which impose fiscal restraint and mandate greater transparency and public accountability; and, government's difficulties in implementing fiscal consolidation despite high levels of debt.

The Impact on Debt of Government's Role as Employer of the Last Resort

In a majority of SIDS, the wage bill is the largest component of public expenditure. Given the large size of government in small states, government is frequently the largest employer in the economy. In the Caribbean, public sector wage bills on average, consume more than 40 percent of total current expenditure. Large-scale employment and the high wage bill have led to deteriorating primary balances and, in turn, significant debt accumulation in SIDS.

Efforts to reduce public sector employment have been explicitly or covertly resisted by the government in many SIDS. In an International Monetary Fund (IMF) discussion with Eastern Caribbean Currency Union (ECCU) member states, the Antigua and Barbuda authorities accurately summarized the common view of the role of government in SIDS as "employer of the last resort" (IMF 2009). These member states maintained that it was critical to prioritize employment in the wake of the 2008–2009 crisis to "mitigate hardships" and to preserve social (and political) stability. Significantly, in Antigua and Barbuda, more than 40 percent of the labor force were public sector employees. An overriding concern of governments is that, given a small and ostensibly weak private sector, there is little scope for surplus labor to be absorbed elsewhere, so that retrenchment of employees is neither socially or politically feasible. They argue further that given the large size of the public sector, downsizing the public sector with little scope for alternative employment would be inimical to growth given decreasing consumption and productivity.

The acquiescence to high wage demands, particularly when powerful public sector groupings (teachers, security, and nurses) have agitated for large increases close to elections, has exacerbated the wage bill problem and imposed an even heavier fiscal burden on governments. In the wake of increasingly unsustainable debts, a number of SIDS governments have faced immense pressure from international financial institutions to implement measures to reduce their fiscal deficits. Few SIDS have implemented the politically difficult task of job cuts when faced with the requirement to reduce fiscal expenditures. In lieu of retrenchment, governments have preferred to engage in social partnerships with public sector employees to keep wages kept low or temporarily frozen so as to preserve jobs. The cost of low salaries, however, has been a flight of quality from the public service. This has affected the overall productivity of the public service and the quality of technical outputs, while locking in high and growing wage costs.

The Impact on Debt of Government as Engine of Growth

With weak private sectors, governments in SIDS tend to assume a dominant role as engine of growth in their economies. SIDS have often sought to implement large scale capital projects, often funded by debt, as the primary means for stimulating growth and promoting economic development. In the absence of fiscal rules, a weak public financial management framework and a clear debt management policy, the cost of project implementation and the impact on the fiscal budget and public debt are often not accorded a high political priority. In the Seychelles, for example, a major thrust to double per capita GDP by 2017, led to the government implementing huge infrastructure projects financed with debt and at huge cost to the fiscal budget. This, along with other macroeconomic imbalances, led to the Seychelles defaulting on its external debt in 2008. The cases of Tonga and Grenada (Box 8.1) clearly illustrate the real or potential fiscal cost of political decisions on the public debt.

Governments have not been averse to establishing large-sized public bodies to both provide public services as well as to implement large scale capital projects. State owned enterprises, however, have invariably been an added source of cost to the fiscal budget and have facilitated large increases in the public debt. The swathe of public corporations has often been driven by a political perspective of "social entitlement to basic needs at minimal cost" or by a "perceived imperative to support the private sector or to fill a vacuum created by the retreat (or absence) of the private sector" (Caribbean Development Bank 2013).

Public enterprises in SIDS have generally been costly to governments. Despite higher salaries to attract expertise and improve institutional capability, most public corporations operate inefficiently. Social and political objectives are preeminent at the cost of economic profitability. The losses

Box 8.1 Political dynamics, the absence of rules, and debt accumulation—two SIDS case studies

In 2007, the government of Tonga had total public debt to GDP of 27 percent. By 2009, Tonga's external debt to GDP had almost doubled, rising to 51 percent of GDP, a consequence of the authorities borrowing two loans from China to fund infrastructure reconstruction projects following riots and property destruction in 2006. The two loans amounted to over 30 percent of Tonga's GDP. Not only was the accumulation of debt large and rapid but, in addition, the two loans, denominated in Chinese Yuan, exposed the government's debt portfolio to significant exchange risk. The depreciation of the Tongan pa'anga against the Chinese Yuan added further to the fiscal burden of the debt. Shortly after the contracting of these loans, the International Monetary Fund in its assessment of Tonga's debt sustainability placed Tonga at "high risk of debt distress" (International Monetary Fund 2010) and proceeded to recommend that "further borrowing by the government, even on concessional terms, should be avoided."

Similarly, in 2010, the government of Grenada sought to borrow a US$107 million loan from China, equivalent to 17 percent of GDP to finance the construction of a 100-room luxury hotel in the island (Caribbean360 2010). The authorities were strongly advised by the International Monetary Fund to rethink contracting the loan because it would jeopardize the country's debt sustainability. Similar to Tonga, the Fund assessed Grenada at a high risk of debt distress. The Fund further intervened by imposing a new criterion under Grenada's 2006–2010 Fund supported program—a ceiling on bilateral external concessional borrowing (International Monetary Fund 2014)—in order to discourage the country from borrowing and adding to its already unsustainable debt burden. At the time, Grenada's public debt was already unsustainable at a 102 percent of GDP.

In retrospect, the Tongan authorities acknowledged that given existing macroeconomic conditions, borrowing such large loans in a small and highly vulnerable economy was imprudent. Both the authorities and the Fund agreed that the absence of a sound debt policy and a comprehensive debt management strategy in line with a medium-term fiscal framework constrained Tonga's ability to make sound borrowings decisions and contributed to the massive debt build up (International Monetary Fund 2010). Notably, it was only in the wake of the severe fiscal constraints, that a debt

(Cont.)

(Cont.)

sustainability policy, which included annual limits for public debt to GDP, was implemented as a means of ensuring that future borrowing decisions would not erode the sustainability of public debt. In Grenada's case, "weak debt management" was cited by the Fund (International Monetary Fund 2014) as a continuing source of macro-fiscal vulnerability.

of enterprises are assumed on the government's balance sheet, eroding fiscal balances and adding to government indebtedness. Even when such entities are privatized or dismantled, the debts that they amass are transferred to the government books, leading to the legacy of increased expenditure obligations.

The Challenge of Fiscal Transparency and Accountability

Contingent liabilities and off-budget items have contributed to significant debt accumulation in SIDS. Yet, they have remained a useful weapon in many governments' arsenal to obscure the full extent of the fiscal burden. There is general consensus that loan guarantees help to provide relatively low-cost funds to beneficiaries. The strong financial backing of government allows public corporations and other entities to negotiate better terms that they otherwise would have received. Since a default by the beneficiary triggers a call on the guarantor—the government—to repay the obligation, such guarantees need to be issued against the careful scrutiny of the beneficiary's balance sheet and an assurance that the beneficiary has the current and future capacity to repay the obligation. Yet governments frequently issue loan guarantees to loss-making enterprises. This suggests that governments are aware of their incapacity to honor their debt obligations. In effect, the government knows, ex ante, that in the future it will assume the obligation and that it, de facto, has entered into a deferred financing arrangement. From a political perspective, this practice has two benefits. It advances the government's public agenda by providing public goods and services and, at the same time, it hides the true fiscal burden of the debt. Government avoids being held to account either by Parliament or by a discerning public.

Deferred financing arrangements and public–private partnerships also conceal the full extent of government's commitments and fiscal burden. Used frequently in Jamaica in the 1990s as a means of financing road infrastructure, deferred financing arrangements allowed for private contractors to use their own resources or debt financing to undertake projects on behalf of the government. At the completion of the project, the government assumed ownership as well as the debts incurred during the undertaking of the project. The expenditures associated with the project would be excluded from budgetary reporting while the government would enter into contractual commitment to repay the contractor. Over time, the bringing to book of these deferred financing arrangements caused a serious deterioration in the fiscal deficit and increased Jamaica's domestic indebtedness. The deferred financing arrangements provided a conduit to bypass Parliamentary approval. In a newspaper article headlined "Ministry Breach—$19 billion in Unapproved Loans," a Ministry of Finance official when sitting before the Public Accounts Committee was quoted as readily admitting that "we recognize the breach [of the law]" (Jamaica Observer 2005).

The often adversarial and "winner takes all" politics has not encouraged governments in SIDS to be transparent or to be held more accountable for fiscal decisions. The hierarchical nature of the political structures, a dominant Cabinet and lopsided parliamentary majorities, reduce the likelihood and/or effectiveness of public and parliamentary scrutiny. This is important, as such scrutiny is needed to identify and challenge practices such as: sloppy or absent procurement procedures; cronyism when allocating positions in "well endowed" state enterprises; massive cost overruns of projects; and the back door deals that characterize off-budget financial arrangements. In such an environment, fiscal reforms that seek to increase transparency and accountability are strongly resisted.

The Political Challenge of Fiscal Consolidation

Deteriorating fiscal deficits and high levels of public debt have forced a number of SIDS to contemplate significant fiscal adjustment. Fiscal adjustment would demand an appropriate mix of expenditure cuts and revenue enhancing measures. However, given the large size of public sector in SIDS, the weight of fiscal adjustment would necessarily fall on the expenditure side (Amo-Yartey et al. 2012). The challenge for SIDS has been how to implement fiscal consolidation measures given a desire to remain politically and economically dominant and to retain their position as "employer of last resort."

In SIDS such as Jamaica, Grenada, and Antigua and Barbuda, all countries which have restructured and remain debt distressed, there is little room for maneuver. Fiscal adjustment is impeded by a high level of fiscal rigidity given the main expenditure components—interest payments, wages, and social security. Yet, the scope for tax increases in some SIDS is also severely limited given already high levels of tax collection.

On the expenditure side, the main scope for fiscal adjustment outside of debt restructuring is to reduce the wage bill. However, as suggested in the previous section, governments baulk at contemplating such a measure. Consequently, attempts at public sector reform have been half-hearted. Given this political dilemma, many governments have opted to significantly cut capital expenditures thereby reducing the medium to long-term scope for economic growth. In a recent study of the fiscal consolidation performance of Caribbean small states, a 2012 IMF study noted the following, "The quality of past fiscal consolidation in the Caribbean is questionable, as capital expenditure has been cut heavily, creating infrastructure gaps inimical to growth and hence compromises the achievement of debt sustainability."

Added to the quality of fiscal consolidation is the question of the political commitment in SIDS to fiscal consolidation and fiscal reform. Undertaking fiscal adjustment measures and implementing fiscal reforms, requires "buy-in" and strong political leadership at the highest levels. Repeatedly governments have faltered in the face of strong resistance and agitation from interest groups and civil society especially in the waning years of an administration, when political survival instincts are at their most engaged.

Changing the Political Agenda (Who Will Bell the Cat?)

The debt problem in SIDS requires a new political agenda that is consistent with sound macroeconomic management. Given the oftentimes natural divergence between political and economic objectives, legislated rules may be the only viable means of ensuring long-term commitment to sound macroeconomic, fiscal, and debt policy. Implementing institutional reforms may also require legislation to ensure that these reforms are entrenched over the long term. While even legislation can be breached, with greater transparency and public accountability, a more aware and educated civil society may hold their governments to a higher standard of fiscal rectitude knowing that this will not leave them saddled with debt for years to come.

To date, institutional reforms have come on the back of debt restructuring program and engaging with the international financial institutions, particularly with the IMF for financial support. While sound fiscal policy by itself is insufficient to address the debt problem in SIDS and must be accompanied by measures to increase growth, weak fiscal policies and the absence of controls is a recipe for continued fiscal distress.

The Need for Strong Institutional Arrangements for Public Debt Management

A weak institutional framework undermines governments' ability to effectively manage its debt. In managing government debt, government must ensure sound fiscal policies to manage the level of debt and the rate of debt accumulation as well as sound debt management to manage the composition of the debt and the attendant costs and risks associated with borrowing. On the fiscal side, government revenues and expenditures must be managed to ensure consistency with debt sustainability. In terms of debt management, governments must avoid risky debt structures, arising from large volumes of foreign currency, variable rate, short-term debt, or contingent liabilities that can lead to increased levels of debt servicing and further debt accumulation in the face of adverse market conditions. A weak legal framework that allows for unfettered borrowing and the amassing of large contingent liabilities; the absence of a clear debt management objective or implementation of debt strategy can lead to poorly structured debt. This, in turn, can cause serious fiscal slippage and further debt accumulation as government budgets are exposed to large and unanticipated increases in debt service costs.

Small island states, to varying degrees, have taken steps to strengthen their institutional framework for public debt management and have met with varying degrees of success. In many instances, the implementation of institutional reforms and the intensity of the effort have been prompted by debt crises in the implementing country. Jamaica, Seychelles, and St. Kitts and Nevis have only undertaken comprehensive institutional reforms in the late 2000s/early 2010s following major restructuring exercises and under the aegis of IMF-supported economic program. Similarly, Barbados' and Grenada's establishment of a debt management unit, Tonga's development of a debt sustainability policy, and the formalization and adoption of binding debt targets among members of the Eastern Caribbean Currency Union were prompted by the rapid pace of debt accumulation in these countries. The reforms are in their early days and it is yet too early to say what impact they have had, if any, on public debt management in these SIDS,

Countries with relatively low debt levels, such as the Pacific island countries (PICS), or those not yet at the brink of a crisis, appear to be less motivated to strengthen their institutional structures. For such countries, progress in establishing a clear debt management objective based on cost-risk considerations, a sound legal framework, a comprehensive debt management strategy, and an efficient and effective organizational structure has generally been less advanced.

As will be explored further, many SIDS face challenges or are reluctant to implement institutional reforms. Where challenges exist, resource constraints associated with small size are frequently a contributing factor. Yet, at the same time, debt accumulation in small states can be traced to weaknesses in institutional structures. A World Bank Study of the debt management performance in 17 small states (Prasad, Pollock, and Li 2013) noted that small states lag significantly behind other developing countries in most dimensions of debt management performance. The point was emphasized that poor debt management practices characterized by excessive or indiscriminate borrowing can "deter new foreign and domestic investment, comprise reform program, depress GDP growth and exacerbate the challenge of meeting debt service obligations." Moreover, poorly structured debt can factor heavily in inducing or propagating an economic crisis. For small states then, there should be an even greater imperative to strengthen institutional structures for debt management given their increased vulnerability to external shocks and natural disasters, their resource constraints and lack of diversification, which, as described in other chapters, has contributed to the exacerbation of debt problems in these countries.

The Legal Framework for Public Debt Management

Sound debt management requires a legal framework that clearly defines the government's authority to borrow, to issue debt, and to undertake debt transactions on the government's behalf (International Monetary Fund and The World Bank 2001). Such a framework should be aimed at limiting potential abuses of power, reducing multiple issuers of debt, ensuring sound financial policies, and establishing clear accountabilities for managing the government's debt (Wheeler 2004).

Perhaps most importantly, modern debt legislation provides for additional accountabilities in executing public debt management. Foremost among these is the requirement by law that the government clearly specifies its debt management objective, expressed in terms of its cost-risk preference. In addition, modern debt management legislation includes provisions that require the government to develop a debt strategy and makes mandatory the requirement that the government reports on its debt management performance against its stated debt management objectives and targets. Further entrenching government accountability, debt management legislation is also expected to explicitly provide for an external audit of debt management operations.

Loan guarantees pose a significant risk to a government's balance sheet. In the event that a beneficiary of a loan guarantee defaults on their payment, the government is contractually obliged to assume the obligation. This can add significantly to the fiscal burden of the debt. Public debt legislation often includes imposing quantitative limits on loan guarantees as an effective means of controlling the growth in these liabilities. Where such provisions are not legislated, SIDS run the risk of rising levels of loan guarantees and incurring heavy costs should such guarantees be called.

Strengthening the Legal Framework for Public Debt Management—The Experience in SIDS

Few SIDS have modernized their public debt management laws (table 8.1). At the end of 2013, the most progress had been made among African SIDS, notably, Mauritius and the Seychelles, both of which had consolidated their debt management laws into a single integrated debt law enacted in 2008. The least progress has been among Pacific island states where debt management laws have yet to be modernized in conformity with international best practice guidelines. In the Caribbean, Jamaica has been the only SIDS[1] to modernize its debt management legislation, doing so in 2012 with the passing of the Public Debt Management Act (2012).

The public debt laws in all three of these SIDS—Jamaica, Mauritius, and Seychelles, share common elements. The laws all specify a debt management objective and require the formulation and implementation of a debt management strategy. Jamaica's Public Management Act (2012) provides that the Minister of Finance shall manage the public debt efficiently and effectively so as to ensure that "the financing needs and the payment obligations of the Government are met at the lowest possible cost over the medium term so as to achieve the fiscal targets specified"

Table 8.1 Commonwealth SIDS: Legal framework by region as at end 2013

	Integrated Public Debt Law	Multiple Laws governing Public Debt
Africa	Mauritius Seychelles	
Asia-Pacific		Fiji Papua New Guinea Samoa Solomon Islands Tonga Tuvalu Vanuatu
Caribbean	Jamaica	Antigua and Barbuda The Bahamas Barbados Dominica Grenada St. Lucia St. Vincent and the Grenadines St. Kitts and Nevis Trinidad and Tobago

Source(s): IMF, World Economic Outlook Database (2014b); World Bank (2009, 2010); National authorities.

(Government of Jamaica 2012). Mauritius' debt management objective is to "meet the borrowing needs of the Government in a manner that avoids market disruption; to minimize the cost of the debt portfolio within an acceptable level of risk; and to support the development of a well functioning market for government securities" (Government of Mauritius 2008). Notably, Seychelles' public debt act does not explicitly specify a debt management objective. However, the law provides that the authority of the Minister of Finance to manage the public debt shall be exercised "in line with the debt management objectives of the Debt Strategy" (Government of Seychelles 2008).

Similar to Seychelles, the public debt laws in both Jamaica and Mauritius mandate the development and implementation of a debt management strategy. The overarching emphasis in both Public Debt Acts is managing and controlling the risk exposures in the debt portfolio. Mauritius' debt legislation states that the debt management strategy shall set risk control benchmarks and medium-term targets for the "composition, currency mix, interest rate mix, maturity profile and relative size of the public sector debt." Jamaica's debt law mandates the government to quantify and analyze the cost-risk trade-off under different debt strategies and economic and financial conditions and establish strategic targets for various risk exposures. The emphasis on risk is not surprising. The debt crises in Seychelles and Jamaica, caused primarily by risky debt structures, were no doubt persuasive in prompting these governments to give weight to risk management in their legislation.

The public debt management laws in Jamaica, Mauritius, and Seychelles also include several other key provisions consistent with a sound legal framework for public debt management. The laws in all three states specify clear purposes for borrowing—to fund the budget and, as specifically stated by Mauritius, to fund investments. The public debt laws in all three SIDS also include quantitative limits on overall debt levels or on explicit contingent liabilities (loan guarantees). In the case of Jamaica, specific ceilings are established on loan guarantees[2] while in Mauritius the law provides that public sector debt shall not exceed 60 percent of gross domestic product (GDP) for any fiscal year.

The debt laws in these three states also explicitly recognize the need to strengthen institutional coordination for public debt management. They therefore provide for the establishment of a high-level public debt management committee, for the expressed purpose of ensuring policy coordination among the principal institutions responsible for public debt management and broad oversight of debt management operations and debt strategy formulation.

The laws also mandate the comprehensive reporting on the public debt to parliament and require an evaluation of debt management performance. Mauritius' public debt law specifies that "the Ministry of Finance shall, not later than one month after the end of every quarter, prepare a report on the outstanding stock of public sector debt, its size and currency composition, interest rate-mix and maturity profile."[3] In addition, it obligates the Ministry of Finance to report on and make public its debt management strategy. Jamaica's Public Debt Act (2012) obligates its government similarly.

It may be too early to assess whether the modernization of the public debt in these three SIDS has led to less debt accumulation. Jamaica's law reform occurred only in 2012 and in the wake of a major debt restructuring exercise and a stringent economic and fiscal reform program supported by the IMF. However, the law has obligated the authorities in Jamaica as well as in Mauritius and Seychelles to focus not only on the level of debt and debt servicing but to more carefully examine and determine the composition of the debt to avoid the debt distress associated with risky structures.

Mauritius and the Seychelles both reformed their debt legislation in 2008. Since the reform, debt to GDP in the Seychelles has fallen by more than half from 130 percent at end 2008 to 62 percent. In contrast, debt to GDP in Mauritius increased from 44 percent of GDP in 2008 to 53 percent at end 2013. However, Mauritius' debt is on a downward trajectory. The 2013 Article IV consultation staff report made the following observation: "Both total public and external debt are on sustainable trajectories and the results of stress tests indicate that debt dynamics are resilient to most shocks." It further stated that the government's fiscal program was "compatible with reducing total public debt to below 50 percent of GDP by 2018" (IMF 2013).

The Challenges of a Weak Legal Framework

While Jamaica, Mauritius, and Seychelles have progressed to a single integrated public debt law, other small states have not proceeded apace. In these SIDS, several Acts govern various aspects of borrowing and other debt management functions. Typically, there is little coherence among the various Acts.

In Trinidad and Tobago, there are almost 20 different laws governing borrowings. These include individual acts that apply to foreign borrowings from various international and regional financial institutions—The World Bank, the Inter-American Development Bank (IADB), and the Caribbean Development Bank (CDB) as well as to the issuance of specific securities, such as Treasury bills, Treasury notes, and savings bonds. Laws governing borrowing and public debt management are similarly dispersed in all the micro states of the Eastern Caribbean Currency Union (ECCU) and the Pacific.

Roy and Williams (2010) argue that there are several benefits to consolidating the array of laws into a single integrated public debt management law. First it supports reforms to improve the focus and professionalism of public debt management. Second, a "coherent and complete act also makes it easier for auditors and parliament to monitor compliance." Finally, an integrated debt law better facilitates transparency and accountability as typically they include a requirement to publish a debt management strategy, report on debt management activities, and publish data outside of the financial statements often unreadable to the general public.

In addition to the numerous laws governing public debt, existing debt management legislation in the majority of SIDS, meets only some of the requirements of a sound legal framework. Debt legislation in virtually all SIDS delegates the authority to borrow and to grant guarantees to statutory bodies to the Minister of Finance. Borrowing limits for new loans and often for guarantees over a specified period are also usually legislated. Frequently, parliamentary approval is legally prescribed for certain types of borrowing and the legislation may also assign the Minister of Finance sole responsibility for managing the public debt. However, absent from debt management legislation in these SIDS are requirements for increased transparency and accountability in debt management operations. These include the requirement to: establish clear, high-level debt management objectives against which government's performance can be evaluated; formulate and implement a debt management strategy based on the high-level objective; report to parliament on debt management operations and in particular, the government's performance in executing the strategy; and, finally to ensure that debt management operations are externally audited. A weak legislative framework provides greater opportunities for imprudent borrowing, especially where, in the absence of an accountability framework, governments can hide the true level of debt by underreporting the level of arrears, off-budget expenditures or level of loan guarantees. In the absence of legislated restraint, borrowing can continue unfettered without the necessary adjustments, leading to further debt accumulation and over time increasing the unsustainability of the debt.

An added problem for many SIDS is that existing debt management legislation is outdated. In the case of Barbados, for example, laws governing treasury securities include archaic conventions for calculating interest. Under existing provisions, interest payments on securities were calculated only on the first day of the month regardless of the issuance and settlement date. Such legal provisions act as a brake on capital market development and no doubt increase the cost of public debt as investors then seek to recoup the potential returns by seeking higher rates of interest. Such provisions have also limited participation in market offers with investors reluctant to participate in issues closing after the first of the month. Less competitive issues increase the cost of government debt and thus the constraints of the law become a factor in adding to the government's debt burden. As debt markets have become more sophisticated, laws in many SIDS have not kept pace. Jamaica, for example, until 2010 could not issue government securities in paperless (dematerialized) form because the then existing laws prior to reform required the issue of paper securities. Thus despite the relatively large size of its domestic securities market, the promotion of a viable secondary market for securities was inhibited as significant delays attended the issue and transfer of securities in paper form.

In SIDS, the absence of modern debt legislation which prescribes that governments specify a clear debt management objective, formalize a debt management strategy, establish debt thresholds, and report on debt management performance, significantly undermines transparency and accountability and weakens debt management. Without this framework, SIDS have considerable opportunity to borrow imprudently and without appropriate accountability. Without attention to the level and composition of the debt, without establishing strategic benchmarks for borrowing, and without a clear understanding of the attendant portfolio risks, there are undesirable incentives to borrow heavily or riskily. This is a prescriptive for increased debt accumulation.

The case of Tonga, as earlier cited, is illustrative. In the absence of any specific reference to the portfolio and to the threats to debt sustainability, Tonga overborrowed. The increases in debt occasioned by the additional borrowing were enlarged by borrowing in foreign currency debt that left the Tongan government heavily exposed to adverse currency movements against the local currency. A sound legal framework with entrenched debt thresholds and a mandated debt strategy may have avoided the surge in indebtedness.

Public debt management becomes opaque when there is no requirement for government to report on its compliance with its stated debt strategy. While legislation in many SIDS provides for the reporting of annual estimates and a financial statement of performance and accounts, far fewer require the government to outline the parameters of the debt management strategy, the agreed measures and strategic benchmarks, and the performance outturn. Thus, without referring to the impact on the debt profile, governments may incur debt obligations without due consideration to their impact on the currency composition, interest rate structure, and maturity profile of the debt.

SIDS face additional pitfalls by having a multiplicity of debt laws governing individual borrowing and this is exacerbated in circumstances where the law prescribes that parliament must approve individual borrowings. Indeed, technical assistance providers and debt management experts strongly caution against parliamentary control extending to individual borrowing decisions. Roy and Williams (2010) emphasize that "parliamentary involvement adds a potentially cumbersome, time-consuming and over-politicized step in the decision making process when time is often of the essence for market borrowing." In SIDS, such as St. Lucia, the requirement to approve individual transactions rather than a global borrowing program adds about 6–8 weeks from the time of decision to borrow and the authority to borrow. Such time lags often close windows of opportunity for low-cost borrowing forcing governments to incur higher than anticipated borrowing costs. Modern legislation overcomes this by seeking to clearly delineate the role of parliament which determines debt policy and approves the debt strategy and the debt managers (technocrats) who are responsible for its execution.

While debt management legislation enables stronger debt management, onerous legal requirements can lead to increased borrowing costs. For instance, limits on individual borrowing instruments may increase borrowing costs and translate into larger-than-anticipated fiscal outlays necessitating even more borrowing. Limits, for example, on certain long-term external borrowings may push government to borrow shorter term in the domestic market, often relying on Treasury bills. This increases refinancing risk and increases debt service costs over the long-term especially if interest as well as principal is rolled over. Similarly, if limits on concessional loans are reached, governments may seek to borrow externally in commercial markets, even if costs are higher, given their limited room to borrow in appropriate markets as determined by the portfolio composition and market conditions. SIDS are replete with examples of borrowing in currencies or in maturities that increase costs and risks having arrived at legally imposed ceilings on certain categories of debt. Wheeler (2004) points out that where legal requirements for long maturities are burdensome in relation to short maturity borrowing, countries will disproportionately rely on the latter, heightening rollover risk in the debt portfolio.

"Good debt legislation underpins sound debt management" (Roy and Williams 2010). While SIDS recognize the need for stronger debt management, reforms to legislation have been slow. The precept that "progress is good as long as it does not involve change" (Wheelan 2010) may describe the approach adopted by SIDS in reforming their debt legislation. Changing legislation is often a lengthy process which governments may find costly. Reforming legislation may also increase the demand for transparency and accountability, requirements that fiscally indisciplined and imprudent governments may wish to avoid. The experiences of the few small states (Jamaica, Seychelles, and Mauritius) that have implemented debt legislation reform suggest that such reforms are most likely to occur in the event of a crisis and often only when pushed to do so by external institutions such as the IMF. For example, Jamaica's enactment of the Public Debt Act (2013) and establishment of a fiscal responsibility framework were required prior actions to secure funding from the IMF under the 2010 Stand By program. Yet, given the high levels of indebtedness among SIDs, it is imprudent to wait until there is a crisis. Reforms to the legal framework are integral in the process of strengthening debt management and removing some of the impediments to achieving debt sustainability.

Debt Strategy Formulation

Developing and executing a debt management strategy is at the heart of public debt management. Public debt management is defined as the process of developing and implementing a medium-term debt strategy that satisfies the government's financing requirements, meets its payments obligations in a manner that minimizes cost and mitigates portfolio risk (Wheeler 2004). Such objectives are likely to be diluted or dissipate altogether in the absence of a debt strategy leading to further debt accumulation.

An explicit debt management strategy sets out how a government will manage the composition of the debt over the medium-term. It will be framed around government's debt policy of either subordinating cost savings to risk reduction or vice versa. If governments have no explicit debt strategy and are unperturbed about risk reduction, then borrowing decisions may focus on short-term cost savings, so as to alleviate immediate budgetary pressures. However, such an approach leaves government budgets exposed to market shocks which, if the exposures are large enough and the shock is severe, can trigger a debt crisis.

The main precursor to developing a debt strategy is establishing a clear debt management objective. The need for a clearly specified objective is twofold. First, the objective anchors the government's debt strategy and provides a clear framework for debt management operations. Second, it concentrates governments' attention on risk as well as cost considerations. Governments when making borrowing decisions must assess not only the cost but the risks and the associated shocks to the debt portfolio which could increase debt service costs over the long term.

Debt management proponents strongly argue that greater emphasis should be placed on risk reduction when the government's debt portfolio is large in relation to the economy's output and includes considerable foreign currency exposure and short-term debt (Wheeler 2004). This is particularly applicable to SIDS. Proponents argue further that countries that have less capacity to manage market risk and that need to borrow in foreign currencies should maintain lower debt levels than countries that can borrow extensively in their domestic currencies. For the former group, minimizing cost should not be the focus of debt management. Instead the emphasis should be on reducing refinancing and market risk.

The debt portfolio composition of many SIDS reflects high levels of currency and refinancing risk. Given a high reliance on external financing, most SIDS have high exposures to foreign currency risk. This is most pronounced in Pacific island states where debt is predominantly denominated in foreign currency. Several Caribbean states, including Dominica, Grenada, St Lucia, and St Vincent and the Grenadines, also have high levels of foreign currency risk reflecting their high shares of external debt. The ECCU countries to some extent have been insulated from adverse currency movements largely because member states maintain a fixed exchange rate pegged to the US dollar. Since foreign currency debts have primarily been denominated in US dollars, ECCU member states have not faced significant currency risk. Moreover, with foreign exchange reserves and exports denominated primarily in US dollars there is a natural hedge to US foreign debt exposures. However, exchange risks are increasing as ECCU member states become more reliant on funding from China with debts mainly denominated in Chinese Yuan. In addition, increasing multilateral support from the European Union has also led to rising shares of Euros in the currency composition of member state debt. Both these developments have led to increased risk exposures that could significantly increase fiscal outlays and add to debt levels in the event of adverse currency changes. SIDS also suffer from high refinancing risk, symptomatic of the shallow domestic debt markets that characterize these states. Much domestic currency borrowing is short-term and denominated by Treasury bills or short-term loans and overdrafts from the banking sector.

Yet, despite high debt burdens and risky portfolio structures, up until 2010 and onward, few SIDS have had an explicit debt management objective and a formalized medium-term debt strategy seeking to minimize these cost and risk exposures (table 8.2). A 2013 World Bank study of the debt management performance in small states found that among the 17 small states sampled, 10 did not formulate and publish a debt management strategy. Of the seven small states that did, only four had a proper decision-making process for developing the debt management strategy and just two met the minimum quality requirements for a debt management strategy (Prasad, Pollock, and Li 2013).

The difficulties in clarifying their debt management objectives and developing debt management strategies can be linked, in good part, to the weak institutional arrangements for public debt management. Dispersed debt management functions across several institutions combined with

Table 8.2 Commonwealth SIDS: Published medium-term debt management strategy (end 2013)

	Debt Management Strategy	No Debt Management Strategy
Africa	Mauritius	
	Seychelles	
Asia-Pacific	Maldives	Tonga
	Fiji	Tuvalu
	Samoa	Vanuatu
	Solomon Islands	
	Fiji	
	Papua New Guinea	
Caribbean	Dominica	The Bahamas
	Antigua and Barbuda	St. Lucia
	Dominica	St. Vincent and the Grenadines
	Jamaica	Trinidad and Tobago
	St. Kitts and Nevis	

Source(s): World Bank (2009, 2010); National authorities.

partial, incomplete, or parallel databases make it difficult for SIDS to have consolidated reports and a complete picture of their debt portfolio so as to develop a comprehensive debt management strategy. Limited staff and weak technical capacity in debt management units are also significant factors. Then, in addition, there are the intangibles such as the commitment of governments to a prudent debt strategy and the discipline required to stay within publicly announced strategic targets and benchmarks. Where SIDS' debt levels are low, as in the Pacific and in commodity exporting small states, such as Trinidad and Tobago, there may be a tendency to minimize the need to routinely assess the portfolio and assess and quantify certain risks in the portfolio. Yet quantifying the cost and risks in the debt portfolio and developing a debt strategy within a broader macro-fiscal framework could yield substantial benefits to low and moderately indebted SIDS to ensure that the overall risk profile of the debt remains low and public debt remains sustainable over the long-term.

Among highly indebted SIDS, a majority of which are in the Caribbean, the need for comprehensive debt management strategies and for more rigorous assessments of portfolio risks is even more pressing. Yet less than 50 percent of Caribbean SIDS produce and publish a comprehensive debt management strategy. The four Caribbean small states that have produced and published a comprehensive debt management strategy are those ranked as the most highly indebted countries not only in the region but among all middle-income countries in the world. Notably, except for Jamaica, these debt management strategies have only been developed after the occurrence of major debt restructuring exercises in the mid to late 2000s. Risky debt structures, global financial shocks, imminent or actual defaults, and the loss of creditworthiness accompanying debt restructuring transactions have given a number of these governments pause for thought. IMF supported economic programs have provided the occasion and sometimes the obligation for governments to formulate a debt management strategy.

Progress toward developing debt management strategies is accelerating although at an uneven pace across regions. African SIDS have moved most speedily in formulating and implementing debt management strategies. At the end of 2013, more than half of all Commonwealth Pacific states had formulated and published debt management strategies. Only Tonga, Vanuatu, and Tuvalu have yet to publish their strategies.

Only in the Caribbean has progress moved more slowly toward developing a strategy to achieve a desired portfolio composition based on instrument type, currency composition and maturity and interest structure. This however is changing. A major initiative was implemented in Eastern Caribbean Currency Union member states to strengthen public debt management capability in the region. The main aim was to "improve capacities of ECCU governments to effectively corral their debt to sustainable levels in line with the fiscal targets established by the ECCB Monetary Council." At the end of 2013, half of the six ECCU member states (Antigua and Barbuda, Dominica, St. Kitts and Nevis) had prepared and published debt management strategies while St. Lucia and Grenada were at various stages of strategy formulation.

Among non-ECCU countries, only Jamaica has formulated and published a medium-term debt management strategy. Jamaica first specified its debt management objective and published a comprehensive debt management strategy in 2000. However, the first attempt did not adequately quantify the risks and costs in the portfolio under alternative economic conditions or subject the portfolio to the rigorous stress tests that are now required. Jamaica's debt management strategy was revamped in 2011 after comprehensive debt management reforms, becoming consistent with the more rigorous standards required by international guidelines for strategy formulation. Jamaica first tabled a comprehensive debt strategy in compliance with the provisions of the Public Debt Management Act in 2012.

Institutional Structures for Public Debt Management in SIDS

Sound debt management practice requires effective institutional structures. While institutional arrangements across countries may differ, sound debt management practice requires clear organizational objectives and roles and the coordination and sharing of information (Wheeler 2004). There is broad consensus too that government debt management generally operates more efficiently if responsibility for decision making and implementation is not spread across several government departments. As will be shown, where the institutional arrangements are disorganized and there is a lack of coordination or capability, borrowing decisions are often not restrained by cost or risk considerations. Thus borrowing may neither be consistent with fiscal sustainability nor with risk or cost minimization, leading to higher volumes of debt than is prudent in the short-term and the potential for higher debt service costs and further debt accumulation in the long-term.

Issues of Fragmented Structures

In many SIDS, however, public debt management functions are dispersed across a number of institutions and as a result there is no integrated approach to debt management. Certainly throughout the 2000s, few SIDS had consolidated their debt management functions in a single organizational entity. In Barbados, for example, debt management functions are performed by several institutions. The Central Bank of Barbados issues long-term domestic securities and executes transactions in the international capital markets; the Treasury issues short-term debt securities, while the Investments Division in the Ministry of Finance sources and negotiates external financing from official creditors. Notably each institution maintained a database only on the subgroup of debt for which it was responsible thereby impeding government's ability to obtain a holistic view of the entire debt portfolio upon which to formulate a single coherent debt strategy. Similarly in the Maldives, no single entity was responsible for managing the public debt. A World Bank mission conducting a debt management performance assessment in the Maldives observed that "there was no integrated approach to the management of total central government debt as there is no single entity responsible for the same" (The World Bank 2009). Moreover the mission further noted that the exchange of information between the central bank and the ministry was infrequent save for the "MMA [the central bank] provided weekly Treasury bills reports to MOFT (Treasury and Public Account) for recording and consolidation purposes."

In many SIDS institutional responsibility for debt management is determined by source of funding (domestic or external), creditor type (official or private creditor), or by instrument type (loans or securities). Often the central bank assumes responsibility for domestic securities issuance given its legally entrenched role as the government's fiscal agent. The central bank may execute transactions in the international capital markets. This devolved responsibility to the central bank usually stems from the bank's knowledge, access, and frequent interaction with financial markets both internationally and domestically. However, other institutions may be responsible for sourcing funds from other markets. In many SIDS, the responsibility for sourcing and negotiating loans with external official creditors is frequently the remit of the Ministry of Foreign Affairs, Office of the Prime Minister, or Planning and Investment Ministries.

Uncoordinated debt management operations are one of the pitfalls of such fragmentation. In such an environment, each institution operates as a "silo," implementing a debt strategy that applies to only their segment of the debt portfolio and usually outside the framework of a single debt management objective. In the absence of a clear debt management objective and comprehensive debt strategy, the risks associated with such an environment are high. Governments are frequently exposed to higher levels of debt service and portfolio risks as inconsistencies in cost-risk trade-offs arise in executing transactions. Central banks, preoccupied with their reserve position, may push to secure more external borrowing despite the attendant foreign currency risk, while project units, anxious to secure investments, may obtain funding from donors in currencies that are not held in foreign reserves and are difficult to hedge against. In the mean time, cost considerations may drive Treasuries to borrow short-term or on a floating rate basis, despite the associated refinancing or interest rate risk.

Fragmentation is not only confined to "front office" fund-raising activities. Many SIDS have back office operations that are frequently dispersed across institutions. In many SIDS, institutions involved in debt management may maintain their own databases, even though typically the debt units in the Ministry of Finance may have and use a comprehensive debt recording system to capture debt records. Typically, the Central banks may maintain their own database for domestic debt securities while finance ministries capture data on external debt. The main drawback for governments in SIDS is the inability to have a full and integrated view of their debt portfolio and thus to develop appropriate strategies to manage it prudently.

Reforming Institutional Structures—Issues and Challenges

Several SIDS have instituted reforms to streamline their institutional arrangements. Jamaica consolidated its debt management operations in 1998, when it transferred all its debt management functions to the

Ministry of Finance. Prior to the transfer, both institutions were recording and publishing information on public debt. With both institutions sharing responsibility for public debt management, there were often divergent perspectives in respect of the maturity and interest structure of the debts to be issued. This reflected the particular remits of each institution, with the Ministry of Finance concerned about increasing debt costs and the impact on the fiscal budget, and the central bank concerned with foreign reserve management and price stability. Consolidating the debt management functions led to a marked strengthening of public debt management in Jamaica. In particular there was a fuller understanding of the portfolio composition of the debt; a more structured approach to primary issuance, and greater reliance on market mechanisms for selling securities; and more timely dissemination of data. The consolidation created increased interaction between the Ministry of Finance and financial markets and a better understanding of investor preferences and the demand for securities. Within two years of consolidating the debt management function within the Ministry, the government of Jamaica was able to develop, table in Parliament and publish its first comprehensive debt management strategy.

Jamaica has not been alone in seeking to consolidate its debt management functions. Mauritius has taken considerable steps at institutional reform, as has Samoa, Seychelles, and Papua New Guinea. A new debt management unit was established in Mauritius in 2002 with the aim of consolidating the government's debt management functions. In Papua New Guinea, debt management functions are centralized in a single entity, the Financial Evaluation Division, which performs key debt management functions including developing and implementing the government's debt management strategy, negotiating the terms of all foreign borrowing, preparing the domestic debt issuance program, recording and reporting on all categories of debt as well as managing government on-lending (The World Bank 2010).

Member states of the Eastern Caribbean Currency Union have also more recently sought to strengthen their institutional arrangements. Except for St Vincent and the Grenadines, all member states have debt management units within their respective Ministries of Finance.

Not all SIDS, at least in the short to medium term, are able to consolidate their debt management functions within a single entity or institution. Capacity constraints are generally the cause. Attracting and retaining staff with the requisite expertise is often difficult in SIDS where there is usually a shortage of the requisite skills and public sector wages are low. While certain efficiency gains are lost and the risk arises of an incomplete or inconsistent debt management strategy, SIDS can still manage their debt effectively using the next best alternative. This next best option requires "a high level of coordination and information sharing among the key agencies involved in debt management" (Wheeler 2004) and the clear delineation of institutional roles and responsibilities. Given these conditions for sound debt management, the main vehicle for such coordination and information sharing is a high-level debt coordination and oversight committee.

The "Next Best" Option—The Role of Debt Management Committees in SIDS

Debt management committees are recommended where debt management operations are scattered across institutions. Such committees serve as a virtual debt office. They become the main vehicle for advising government debt policy and strategy. They provide oversight and ensure coordinated operations among the main institutions involved in debt management. Usually chaired by the chief technocrat in the Ministry of Finance (the Permanent Secretary) the members of such a coordinating committee usually include the executive heads of the principal debt management institutions, namely, the central bank, the treasury, and the planning or investment ministry. A technical committee is usually formed to provide support to this high-level committee. Its primary core function is to provide analytical support to the policy committee as well as to address operational issues related to public debt management (Roy and Williams 2010).

A growing number of SIDS have established high-level debt management committees (table 8.3). In 2012, the Solomon Islands formalized its governance framework, a component of which was establishing a high-level debt policy committee—the Debt Management Advisory Committee (DMAC). The Committee's primary mandate is to "assess the volume and risk characteristics of debt" so as to ensure, inter alia, that the public debt is: "sustainable and affordable; below the thresholds established in the Debt Strategy and Annual Borrowing Plan"; and, "the borrowing aligns with the Medium Term Fiscal Strategy" (Ministry of Finance—Solomon Islands 2012).

In 2010 and 2011 respectively, Barbados and Jamaica each established two debt management committees. Barbados established a National Debt Advisory Committee to provide policy advice to the Minister of Finance and to oversee and coordinate all debt management operations and a Technical Debt Management Committee to provide technical support to the high-level committee. Jamaica's two committees—the Public Debt Management Committee and the Public Debt Financing Committee (PDFC)—were established under the provisions of the 2012 Public Debt Management Act. Similar to Barbados and Solomon Islands, the scope

Table 8.3 Commonwealth SIDS: Debt management committees (as at end 2013)

	High-Level Debt Management Committee	No High-Level Debt Management Committee
Very High Debt (Debt/GDP over 90%)	Antigua and Barbuda Barbados Jamaica Grenada	St. Kitts and Nevis
High Debt (Debt/GDP over 60% – 90%)	Seychelles	The Bahamas Dominica Maldives Samoa St. Lucia St. Vincent and the Grenadines
Medium Debt (Debt/GDP over 30% – 60%)	Mauritius Trinidad and Tobago	Fiji Papua New Guinea Tonga Vanuatu
Low to Medium debt (Debt/GDP 0% – 30%)	Solomon Islands	Tuvalu

Source(s): International Monetary Fund (2014b) and National authorities.

of the high-level debt committee is narrowly defined, focusing primarily on the development and implementation of government's debt strategy. The specific role of Jamaica's Public Debt Financing Committee is to provide technical advice on the government's funding strategy. It is also mandated to review and monitor issues related to the cost or pricing of contingent liabilities and to evaluate the funding requests by public bodies (Government of Jamaica 2012).

A survey of Commonwealth SIDS (table 8.2) indicates that Africa and the Caribbean are most advanced in reforming institutional arrangements to address the dispersion of debt management functions. Notably, with the exception of Trinidad and Tobago, all the Caribbean countries that have established debt management committees have experienced a debt crisis and have undertaken a comprehensive restructuring exercise. Among African SIDS, the experience has been more mixed. Seychelles' debt management committees emerged in the aftermath of its 2008 debt restructuring exercise, while Mauritius established its debt management at the same time as Seychelles, but independent of a major debt restructuring exercise. However, the institutional reforms appeared to have been prompted by the worsening fiscal environment faced by Mauritius in the aftermath of the global economic and financial crisis. Among Pacific SIDS, to date, only the Solomon Islands has established a debt management committee. The relatively more indebted Pacific island states of Samoa, Fiji, Papua New Guinea, Tonga and Tuvalu have not yet done so.

The experience among SIDS suggests that, by and large, changes to institutional structures, whether by centralizing debt management functions in a single debt management entity or by creating oversight committees to ensure coordination and strategic direction, can be readily accomplished. However, the main issue remains whether the changes to strengthen debt management and make more prudent borrowing decisions will be instigated by governments in the absence of a debt crisis.

The Organizational Structure and Staff Capacity

The organization of a debt management unit may play a significant role in the effectiveness and efficiency of public debt management. The organizational structure should allow governments to both efficiently and effectively operationalize the government's debt management objective and debt management strategy. A poorly organized unit can lead to weak management of human and financial resources and undermine the internal controls necessary to minimize operational risk.

Modern debt management offices typically are organized into three functional areas, commonly referred to as front, middle, and back offices. The front office secures financing on behalf of the government, the middle office develops a debt management strategy and the back offices records and reports on executed transactions and ensures the timely settlement of debt obligations. This functional organization is the structure adopted in financial institutions (Wheeler 2004) and allows for a better streamlining of debt management functions and management of operational risk.

The separation of functional activities is intended to help minimize operational risk, particularly the risk of fraud (as seen in the case of Kenya, Box 8.2), and avoid the large financial losses that can and have arisen due to an overlap of functional responsibilities. It is these financial losses, inevitably assumed by the government on its balance sheet, that add to the fiscal burden and lead to further debt accumulation.

The principle of segregated functions has been applied to modern debt management offices which, in a growing number of countries, have adopted similar functional structures to those of private financial institutions. In a government debt management office sizeable capital losses can occur from fraudulent activities in the back office which is responsible for maintaining financial records and overseeing or effecting the settlement of debt service payments. This is especially so given the size of a government's debt portfolio—typically the largest in any given country.

Box 8.2 Segregated debt management functions and debt accumulation—the case of Kenya

The well-documented case of the Anglo Leasing scandal in Kenya involving fraud related to contracts amounting to over US$750 million (The World Bank 2007) underscores the need for segregated functions. The lack of segregation led to fraud and the massive accumulation of debt in Kenya. Some 18 contracts were signed by the Kenyan government in the late 1990s and early 2000s with a private financing company, Anglo Leasing and Finance Limited, to provide goods and services through, as it turned out, fictitious companies. Promissory notes were issued to the company. While the goods and services were never delivered and the contract amount showed "massive overpricing" the debt management unit both approved the terms and conditions of the contract and effected payments on the debt. An investigation in 2004 found that the "commitment was made outside the government budgetary process," had not been approved by Parliament, and it was impossible to ascertain how the contract sum was determined." At the height of the scandal, the Minister of Finance, senior Ministry of Finance officials as well as the head of the debt management unit resigned. The Public Accounts Committee cited weak debt management, including the responsibility to check the legality of the contracts and "serious cases of fraud due to the absence of due diligence" as a major factor in the Anglo Leasing affair. More recently, in 2014, a number of companies holding these promissory notes filed suit against the Kenyan government. The courts ruled in favor of the holders of the promissory notes, requiring the Government to pay the holders the full value of the notes. The Kenyan government decided to honor the judgment noting among other things that it was necessary to "protecting Kenya's economy on account of rising interest rates occasioned by domestic borrowing due to constrained access to international borrowing" and to "maintaining and improving Kenya's credit rating currently at B+."

Few SIDS have debt management units organized into front, middle, and back offices and fewer have a middle office at all. The absence of a middle office is a significant challenge to effective public debt management. The difficulty in establishing a middle office to a large extent reflects capacity constraints in SIDS and in the public service in particular. It also is a legacy effect of the initial creation of debt units primarily as a recording and payments unit.

Many debt management units were formed in the 1980s and 1990s in response to the government's need to have a comprehensive picture of their external debt portfolios. Often, the establishment of the unit coincided with a debt payment crisis or imminent debt restructuring exercise and discovery that external debt records were inaccurate or incomplete. The back office function of accounting for and reporting on debt, and monitoring external debt payments was the pre-eminent function. Front-office functions were dispersed across several agencies and the middle-office function was primarily undertaken by research or economic divisions within the central banks or within macro-fiscal unit within finance ministries. This was the pattern of organizational establishment and development in most SIDS across all regions—Africa, Asia, Pacific, and the Caribbean.

For many SIDS, the transition to a more expanded role for the debt management unit has either not occurred or occurred very slowly. Middle-office capability has tended to develop outside of the debt management unit while front office function continue to be performed in central banks, project and planning divisions of finance ministries or on occasion in the Office of the Prime Minister. Debt management units therefore continue to primarily discharge the back-office function with little or no middle-office capability.

The lack of front and middle-office functions in debt management offices has undermined sound public debt management and increased the risk of poorly structured debt and, as a result, further debt accumulation. It is the front office that negotiates and executes transactions in various financial markets. It is the front office which, knowledgeable about market sources and creditor practices, determines the type of instruments (loans or securities) as well as the currencies, maturities, and interest structure in which borrowings should occur, ensuring consistent with the government's stated debt management strategy. The lack of technical capability makes poor borrowing decisions more likely, leading to higher debt service costs or risks being imported into the government's balance sheet. It is these larger than anticipated debt service costs that lead to higher fiscal expenditures and, over time, further increases in the public debt. Notably, in a World Bank study on debt management performance assessment, small states have been found to "score poorly" on key front office functions, namely, evaluating loan terms and determining the most "beneficial and cost effective borrowing terms and conditions" (Prasad, Pollock, and Li 2013).

Staff Capacity

Informed borrowing decisions do not occur by themselves. Governments must rely on the technical capacity available within the civil service and more specifically within those entities responsible for managing public debt. Informed borrowing decisions require, among other things, a clear debt management strategy anchored within a sound macro-fiscal framework, which defines very precise parameters for negotiating and contracting new loans. Such a strategy will be concerned with the markets in which governments borrow and the impact that such borrowings have on the composition of the portfolio and the risk profile of the debt. Developing and implementing a debt strategy and recording and reporting on debt data in a manner that supports rigorous analysis requires highly competent staff.

Effective debt management requires staff with a combination of financial market, economics, and public policy skills (Wheeler 2004). Debt management fails to be implemented successfully where such skills are inadequate.

SIDS are particularly challenged in attracting and retaining skilled debt management staff. That back office functions are the main operational activities of debt management offices is not by chance. Many SIDS opt to recruit less experienced or clerical staff to perform back office operations largely because such staff are in greater supply, more affordable, and less in demand from the private sector than specialist staff with expertise in finance, financial markets, or economics.

Organizing along functional lines requires specialist staff and manpower, both of which are scarce in SIDS. Given their small populations, finding a large pool of any specialized set of skills in SIDS is difficult. Recruiting staff for specialist positions within the front and middle offices of debt management units poses a significant challenge to governments. An added problem is that within small debt management offices there is not enough staff for tight specialization in roles and frequently there is not enough work to justify the appointment and retention of those with very specialized skills. This problem is clearly exemplified in many SIDS where the average staff size of a debt management office is four (table 8.4). Such offices can ill-afford specialists. Moreover a single specialist creates key person risk. This is the risk that only one or two persons have a particular skill thereby severely disrupting work activity when they are temporarily or permanently absent. In SIDS functional areas are frequently disassembled and staff work on whatever is required. While this fully occupies staff, the specialist skills required for formulating and implementing debt management strategies are often not developed and as a consequence, governments increase their risk of imprudent borrowing decisions.

"Debt management reforms often fail because funding for hiring skilled staff is inadequate and inconsistent with stated goals" (Wheeler 2004). Attracting and retaining specialist staff in debt management units is a major challenge faced by SIDS. A debt management performance

Table 8.4 Selected SIDS: Staff size in MoF debt management unit (as at end 2013)

Country	Size of DMO (no of staff)
Africa	
Mauritius	5
Seychelles	5
Asia-Pacific	
Maldives	12
Papua New Guinea	31
Samoa	5
Solomon Islands	4
Tonga	4
Caribbean – ECCU states	
Antigua and Barbuda	6
Dominica	3
Grenada	3
St Kitts and Nevis	2
St. Lucia	5
Caribbean – Non-ECCU states	
Barbados	3
Jamaica	28
Trinidad and Tobago	7

Source(s): National authorities.

assessment of the Maldives found that "Because of the low salaries, as compared with that offered by the private sector and independent commissions, it is difficult to attract experienced personnel to the Ministry. Therefore staff capacity needs to be built by recruiting skilled staff. New intakes often come only with A level education (high school graduates) and require further education and training" (The World Bank 2009).

The Maldives is far from the exception. In 2011, Jamaica embarked on major reforms to strengthen its public debt operations. The organizational structure of the DMU was completely overhauled to enable it to better develop and execute a debt management strategy (Caribbean Development Bank 2013). The debt structure provided for a staff complement of 8 in the middle office. Up until early 2014, the debt management unit had failed to recruit high-level middle-office staff after repeated tries. Low public service salaries were identified as a major impediment to recruitment. A Commonwealth Secretariat study on the reorganization of Jamaica's debt management unit (Commonwealth Secretariat 2010) found that salary differentials between positions in the debt management unit and their equivalent in the private sector were as high as 30 percent and that this value under-estimated the true differential. At the start of 2014, only two-thirds of all positions in the middle office had been filled.

SIDS are devising ways to overcome the challenge of manpower constraints and low salaries. In some instances, such as in the Eastern Caribbean, steps have been taken to increase the technical capacity of staff through targeted training and capacity building. The Canada Eastern Caribbean Debt Management Advisory Services (CANEC-DMAS) is helping to strengthen debt management capacity among member countries. More member states have produced debt management strategies as a result of this intervention. Many SIDS have eschewed employing experts opting instead to approach recent university graduates with offers of significant training opportunities and career path development to compensate for low levels of remuneration. The launch of regional groups such as the Caribbean Association of Debt Managers (CARADEM) in 2011 is another means by which some SIDS have sought to strengthen debt management capacity.

Conclusion

This chapter has explored how political, institutional, and organizational structures may lead to imprudent borrowing decisions in SIDS adding to already high levels of indebtedness. It examined the legal, institutional, and organizational framework for public debt management and sought to illustrate how weak institutional structures undermine effective debt management and often lead to larger than estimated increases in the public debt.

The chapter found that the large size of government in SIDS, combined with their paternalistic approach to governing (particularly their role as employer of last resort and as a "cradle to grave" provider of public goods), has helped to undermine fiscal policy and contribute to higher levels of debt. The chapter found that there are significant gaps between international best practice for public debt management and the common experience in many SIDS. Some major impediments to reform implementation relate to resource constraints especially in relation to staff capacity. Other impediments are more nuanced, and, in part, reflect government reluctance to become more transparent and accountable and to be subject to the discipline and restraint imposed by legislation. Those countries most advanced in implementing reforms were those most likely to have had borrowed excessively and been thrown into a debt crisis. It is in the absence of options and with a strong push from the IMF that progress on implementing reforms has been made.

SIDS have much to gain from strengthening their institutional structures for public financial management. Increased transparency and accountability in fiscal operations as well as the enactment of fiscal rules can serve to increase fiscal discipline. While it is well established that loose fiscal policy and widening fiscal deficits lead to debt accumulation,

SIDS can further increase the level and volatility of the public debt by relying on low-cost rather than low-risk debt. SIDS must implement reforms to allow stronger management of the public debt or risk recurring episodes of high indebtedness and debt distress.

Notes

1. It should be noted that Suriname preceded Jamaica and modernized its debt laws in 2003. However, it is not classified as a SIDS for the purpose of this study.
2. Quantitative ceilings on total public sector debt are provided for in fiscal legislation.
3. Mauritius Public Debt Act (2008) Section 9 (4) and (5).

References

Amo-Yartey, Charles, Machiko Narita, Garth Peron Nicholls, Joel Okwuokei, Alexander Peter, and Therese Turner-Jones. 2012. "The Challenges of Fiscal Consolidation and Debt Reduction in the Caribbean." *IMF Working Paper WP/12/275*. Washington, DC: International Monetary Fund.

Brown, Deryck R. 2009. *Institutional Development and Reform in Public Services: The Experience of Small Caribbean States*. Conference Paper, London: Commonwealth Secretariat.

Caribbean Development Bank. 2013. *Public Sector Debt in the Caribbean: An Agenda for Reduction and Sustainability*. Bridgetown: Caribbean Development Bank.

Caribbean 360. "IMF Tells Grenada to Think Twice About China Loan." Caribbean 360, January 14. http://www.caribbean360.com/business/imf-tells-grenada-to-think-twice-about-china-loan.

Commonwealth Secretariat. 2010. *Restructuring the Debt Management Unit—Jamaica*. Technical Report. London: Commonwealth Secretariat.

Government of Jamaica. 2012. "Public Debt Management Act (2012)." Government of Jamaica.

Government of Mauritius. 2008. "Public Debt Management Act (2008)." Government of Mauritius.

Government of Seychelles. 2008. "Public Debt Act (2008)." Government of Seychelles.

———. 2010. *Debt Management Strategy—2010–2012*. Strategy Report, Victoria: Government of Seychelles.

Haque, Tobias A., David S. Knight, and Dinuk S. Jayasuriya. 2012. *Capacity Constraints and Public Financial Management in Small Pacific Island Countries*. Policy Research Working Paper 6297, Washington, DC: The World Bank.

International Monetary Fund (IMF). 2008. *Big Government, High Debt and Fiscal Adjustment in Small States*. Washington, DC: International Monetary Fund.

———. 2009. *Discussion on Common Policies of ECCU Member Countries.* IMF Country Report No. 09/175, Washington, DC: International Monetary Fund.

———. 2010. *Tonga: IMF Article IV Consultation—May 2010.* IMF Country Report No 10/112, Washington, DC: International Monetary Fund.

———. 2013. *Mauritius: 2013 Article IV Consultation—April 2013.* IMF Country Report No. 13/97, Washington, DC: International Monetary Fund.

———. 2014a. *Grenada—Ex Post Assessment of Longer Term Program Engagement.* IMF Country Report No 14/19, Washington, DC: International Monetary Fund.

———. 2014b. *World Economic Outlook.* Washington, DC: International Monetary Fund.

International Monetary Fund (IMF) and The World Bank. 2001. *Guidelines for Public Debt Management.* Washington, DC: International Monetary Fund.

———. 2003. *Guidelines for Public Debt Management—Accompanying Document and Case Studies.* Washington, DC: International Monetary Fund and The World Bank.

Jamaica Observer. 2005. "Ministry Breach—$19 Billion in Unapproved Loan Guarantees." March 23. http://www.jamaicaobserver.com/news/77397_MINISTRY-BREACH----19-6-billion-in-unapproved-loan-guarantees. Accessed May 28, 2014.

Medina Cas, Stephanie and Ota Rui. 2008. *Big Government, High Debt and Fiscal Adjustment in Small States.* IMF Working Paper WP/08/39. Washington, DC: International Monetary Fund.

Ministry of Finance—Solomon Islands. 2012. *2012 Debt Management Strategy.* Strategy Paper. Honiara: Solomon Islands Government.

Prasad, Abha, Malvina Pollock, and Ying Li. 2013. *Small States—Performance in Public Debt Management.* Policy Research Working Paper 6356. Washington, DC: The World Bank.

Roy, Arindam and Mike Williams. 2010. *Government Debt Management: A Guidance Note on the Legal Framework.* London: Commonwealth Secretariat.

Wheelan, Charles. 2010. *Naked Economics—Undressing the Dismal Science.* New York: W.W. Norton & Company.

Wheeler, Graeme. 2004. *Sound Practice in Government Debt Management.* Washington, DC: World Bank.

World Bank. 2007. *Managing Public Debt: From Diagnostics to Reform Implementation.* Washington, DC: World Bank.

———. 2009. *Debt Management Performance Assessment (DeMPA)—Maldives.* Technical Report, Washington, DC: World Bank.

———. 2010. *Debt Management Performance Assessment (DeMPA)—Papua New Guinea.* DeMPA Report, Washington, DC: World Bank.

Chapter 9

Does Debt Restructuring Work? An Assessment of Remedial Action in SIDS

Michele Robinson

Restructuring of Public Debt: An Overview of Available Mechanisms

Since the 2000s, a growing number of Small Island Developing States (SIDS) have undertaken debt restructuring in the face of high and increasingly unsustainable debt. This chapter examines the current debt restructuring mechanisms available to SIDS and the extent to which using these mechanisms has provided debt relief. The chapter focuses on SIDS that have undertaken comprehensive debt restructuring since the start of the 2000s and assesses whether debt restructuring has met their objective of placing their debt on a downward trajectory so as to restore debt sustainability over the medium term.

The Framework for Restructuring

The unsustainable debt burden of several small island developing states (SIDS) has, for some, escalated into full-blown debt crises requiring comprehensive restructuring of their debt. Under recent debt reorganizations, both domestic and external debt have been restructured involving a range of creditor classes and instrument types. The type of reorganization has reflected the particular portfolio composition of the affected SIDS and the amenability of the debt to restructuring.

Among external debt obligations, official bilateral debts have generally been the most amenable to debt restructuring. Commercial bank debts, while often more difficult to restructure especially if many banks were involved, have also been reorganized. In both cases, the framework for renegotiation—whether under the Paris Club[1] for official bilateral debt or the London Club[2] for commercial bank debt—is well established and the workout mechanisms are well understood.

The restructuring of sovereign bonds has been a more recent development on the global financial landscape. However, given multiple bondholders who are often heterogeneous and widely dispersed, sovereign bond restructuring is generally far more difficult to execute. Debt exchanges have been the main mechanism for restructuring such bonds. Bondholders exchange their existing bonds for new bonds which reflect the terms of the restructuring. However, the ability of a sovereign to undertake an exchange is determined by the law governing the sovereign bond issue. Most sovereign issues are issued under New York law. However, until 2003, sovereign bonds issued under New York law did not include collective action clauses. This prohibited a change in the terms of the existing bond unless there was full consensus among bondholders. However, there has been a major shift in the global financial architecture and by the end of the decade most sovereign issues include collective action clauses. These clauses allow a super-majority of bondholders—often 75 percent—to agree terms which would then be binding on all bondholders. This limits the possibilities of holdouts—bondholders who do not agree to the new terms thus limiting the relief available from debt restructuring.

Since the 2000s, six SIDS—Antigua and Barbuda, Dominica, Grenada, Jamaica, the Seychelles, and St. Kitts and Nevis—have restructured their debts. The success of their debt reorganizations and the amount of relief obtained has varied across countries and, to a great extent, has been determined by the portfolio composition of the debt. While some of these debt-distressed SIDS have seen reduced debt levels and declines in their debt-to-GDP ratios, many still are struggling to reverse the continued debt accumulation. To date, none of these SIDS has seen their public debt return to sustainable levels.

Experiences with Official Bilateral (Paris Club) Creditors

Over the period 2006–2012, four debt distressed SIDS—Antigua and Barbuda, Grenada, Seychelles, and St. Kitts—restructured their official bilateral external debts through the Paris Club (table 9.1). The Caribbean SIDS, Antigua and Barbuda, Grenada, and St. Kitts and Nevis, had their bilateral debt restructured on "Classic" Paris Club terms—the standard and least generous of Paris Club restructuring options. Only Seychelles restructured its Paris Club debts on "Evian" terms, which are typically more generous than Classic terms as creditors take into account the relief required to achieve debt sustainability.

Under Classic terms, only debts falling due over the duration of an International Monetary Fund (IMF) program (1–4 years) are eligible for

Table 9.1 SIDS: Summary of debt restructuring terms from the Paris Club, 2000–2012

Country	Date of Agreement	Consolidated Amount (US$ million)	Consolidation Period (months)	Paris Club Approach	Paris Club – Terms of Restructuring
Antigua and Barbuda	September 16, 2010	110.0	32	Classic	Repayment over 12 years inclusive of 5 years grace.
Grenada	May 12, 2006	16.0	48	Classic	Repayment over 12 years inclusive of 5 years grace.
Seychelles	April 16, 2009	163.0	8	Evian	Stock Reduction of 22.5 percent as at July 1, 2009 and 29.3 percent reduction as at July 23, 2010
St. Kitts and Nevis	May 24, 2012	5.0	26	Classic	Repayment over 20 years inclusive of 7 years grace

Source(s): Paris Club (2009, 2010, 2012). http://www.clubdeparis.org.

rescheduling. These debts are typically rescheduled over 12 years including a grace period. These terms are far less favorable than those accorded to lower middle income or poor countries where debt may be rescheduled for up to 40 years. Moreover, unlike terms extended to highly indebted poor countries, Classic terms do not include debt cancellation.

Antigua and Barbuda and Grenada both received classic Paris Club terms where debts were rescheduled over 12 years inclusive of 5 years grace (Paris Club 2010). The amounts rescheduled represented only a small share of their total external debt with Antigua and Barbuda rescheduling US$110 million or less than one-fifth of its total external debt and Grenada rescheduling US$16 million or around 4 percent of its external debt.

St. Kitts and Nevis also received "Classic" terms received from the Paris Club. However, given its exceptionally high debt burden, with debt to GDP levels exceeding 160 percent, it received longer repayment terms than those received by Antigua and Barbuda, and Grenada. The Paris Club rescheduled debts falling due over a 26-month period for 20 years inclusive of a 7-year grace period (Paris Club 2012). While this provided significant cash relief, amounting to 90 percent of debt service payments due, this only affected some 4 percent of the country's total external debt. Overall, therefore, the relief secured from the Paris Club did not significantly ameliorate the country's debt burden.

The Seychelles has been the only SIDS to obtain debt relief from the Paris Club on "Evian" terms.[3] Under Evian terms, which are extended to countries who do not qualify under the Heavily Indebted Poor Country (HIPC) initiative, the Paris Club explicitly considers a country's debt sustainability and seeks to provide a restructuring solution that helps to restore long-term sustainability and ensures a permanent exit from the Paris Club restructuring. Thus Evian terms include debt reduction as well as debt rescheduling (International Monetary Fund 2004b).

With debt-to-GDP in excess of 150 percent at the end of 2008 and external debt comprising 60 percent of the total, with most in arrears, the Seychelles presented a strong case for debt reduction under "Evian" terms. Under the terms of the 2009 Paris Club agreement, US$163 million of official bilateral debt comprising arrears as at end-October 2008 as well as maturities falling due between November 2008 and June 2009 were consolidated and rescheduled (Paris Club 2009). In July 2009, 22.5 percent of official bilateral debt outstanding was cancelled, with a further 29.03 percent was written off in July 2010. The Paris Club debt reduction was subsequently accompanied by a debt exchange which led to a further reduction in the external debt. Overall public debt-to-GDP dropped from 129 percent in the year prior to the restructuring to 82 percent of GDP in 2009. Debt-to-GDP has remained on a downward trajectory since the debt restructuring operation was concluded.

Experiences with Debt Exchanges and Private Bondholders

Although Paris Club debt restructuring did provide debt relief to some SIDS, it was largely inadequate. Debt exchanges were the main avenue for debt restructuring by SIDS since their debts were owed primarily to private creditors, mainly bondholders. Since the start of the 2000s, five of the six SIDS that restructured their debt, have used debt exchanges to restructure their bonded debt and commercial bank loans. Relief has varied across countries depending on the terms sought and the participation of affected creditors (table 9.2).

Dominica was the first SIDS to undertake a debt exchange during the 2000s. It used this mechanism to restructure commercial loans and bond issues owed to domestic and external creditors. Dominica's 2004 debt exchange was notable in two ways. First, it was the first country to adopt and implement a "cooperative" approach based on an internationally agreed code of conduct to guide restructuring operations involving sovereign debtors and private creditors (International Monetary Fund 2005). Second, the debt exchange faced considerable problems from holdouts and litigation.

Dominica offered to exchange its existing domestic and external private debt for three new bonds; two issued at a discounted face value and all three structured with longer maturities and a reduced interest rate of

3.5 percent. Given the small number of Paris Club creditors, there was no scope for negotiation through the Paris Club and official bilateral creditors were invited to participate in the exchange as well as private creditors (International Monetary Fund 2004a). Dominica also secured debt relief from the Caribbean Development Bank (CDB) as they agreed to significantly lengthen the repayment periods and lower interest rates on affected debts.

While success was achieved in restructuring debt with official creditors, success was more limited among private creditors. After twice extending the closing date for the offer, only a 70 percent participation rate in the debt exchange was achieved—the majority of which were domestic lenders. Holdout creditors proceeded with legal action against the Dominican authorities to maintain payments under the original terms of the instrument. As a "good faith" gesture, the authorities placed the amounts due under the new terms of the offer in an escrow account available to holdouts on acceptance of the new terms. Dominica's debt exchange was not concluded until 2007.

Similar to Dominica, Grenada used a debt exchange in 2005 to restructure its domestic and external commercial obligations. Half of Grenada's public debt or some US$276 million was restructured. External debt of US$190 million, including a US$100 million global bond, as well as US$86 million of domestic bonds and loans, was exchanged for two new instruments—a new US dollar bond and Eastern Caribbean (EC) dollar bond. However, unlike Dominica, there were no principal write-offs but a substantial reduction in present value terms. Both bonds were repayable in 20 years and had stepped-up interest rates. Participation in the exchange was high at 91 percent.

Seychelles confined its debt exchange only to its external debt. Implemented in February 2010, Seychelles debt exchange had a number of landmark features. First, the exchange set a historic precedent when it received a US$10 million guarantee from the African Development Bank (AfDB) which partially covered interest obligations under the exchange offer (African Development Bank 2010). The involvement of the multilateral financial institution provided Seychelles with significant leverage in negotiating with its creditors and in achieving a successful debt exchange. It also complemented the efforts of the IMF and other donors to support the Seychelles' economic reforms and to restore long-term debt sustainability. Second, the exchange also included a "principal reinstatement" clause or "clawback" provision which allowed for an upward adjustment in the principal to be repaid in the event that Seychelles failed to satisfy certain criteria under the IMF program (Robinson 2013). The provisions stipulated that an additional 25 percent of the face value of the new bond would be paid to bondholders if Seychelles failed its first

Table 9.2 SIDS: Summary terms of debt exchange, 2000–2012

	Country	Date of Exchange	Debt Affected	Terms of Exchange
2004	Dominica	September 2004	Bonds (external and domestic); bank loans	• Old debt exchanged for: • New 3.5% short discount bond 2014 (30% discount); • New 3.5% intermediate discount bond 2024 (20% discount); • New long par bond 2034.
2005	Grenada	November 2005	Bonds (external); bank loans (external)	• Old debt exchanged for: • New US$ par bond 2025 • New EC$ par bond 2025 • Coupon rate (stepped-up): 0.85–8.0%
2009	Seychelles	December 2009	Bonds (external and domestic); bank loans	• Old debt exchanged for: • New US$ discount bond 2026 (50% discount). Stepped-up coupon • New 2% US$ par bond 2041 • Partial guarantee from AfDB
2010	Jamaica	January 2010	Bonds (domestic)	• Old bonds exchanged for 24 new benchmark bonds: • New J$ fixed-rate par bonds • New J$ variable-rate par bonds • New US$ fixed-rate par bonds • New CPI-indexed bonds
2012	St. Kitts and Nevis	April 2012	Bonds (external and domestic); bank loans	• Old debt exchanged for: • New US$ discount bond 2032 (50% discount). Stepped-down coupon 6–3%) • New 1.5% EC$ par note 2057 • Partial guarantee from CDB
2013	Jamaica	February 2013	Bonds (domestic)	• Old benchmark bonds exchanged for new benchmark bonds: • New J$ fixed-rate par bonds • New J$ variable-rate par bonds • New 5.25% US$ fixed-rate par bonds • New CPI-indexed bonds

Source(s): Various International Monetary Fund Articles for various years, National authorities.

review of its IMF program. Finally, the Seychelles exchange was the first modern debt exchange not to entertain residual holdout bondholders. Seychelles triggered the collective action clause embedded in the original instruments which allowed a super-majority of original bondholders to apply the new bond terms to the original claim. A 100 percent participation rate was therefore achieved.

Seychelles exchanged US$321 million in external debt, including two global bonds, for new debt securities—a new discount note and a new par note. The new discount note reduced principal by 50 percent and extended maturities until 2026. Similar to Grenada's debt exchange, the discount instrument provided front-loaded relief with a stepped-up interest rate, set initially at 3 percent. The par note was fixed at 2 percent and matured in 2041.

St. Kitts and Nevis also received a partial guarantee under its March 2012 debt exchange. The Caribbean Development Bank guaranteed principal and interest falling due up to a limit of US$12 million on a rolling, reinstatable but non-accelerable basis (International Monetary Fund 2012a). The St. Kitts and Nevis debt exchange restructured US$150 million of external and domestic debt securities and bank loans. The restructured debt accounted for around 10 percent of total public debt. The existing instruments were exchanged for two new instruments—a new US dollar bond which discounted principal by 50 percent and a new EC dollar par bond. The initial interest rate on the new US dollar bond of 6 percent for the first four years was to drop to 3 percent for the remaining life of the instrument (International Monetary Fund 2012b). The local currency par bond had a maturity of 45 years, including 15-years grace and an interest rate of 1.5 percent. Voluntary participation in the debt exchange was high, at 96.8 percent. Like the Seychelles, St. Kitts and Nevis triggered the collective action provisions in the existing instruments to increase participation to 100 percent.

St. Kitts debt restructuring also entailed a debt-for-land swap with domestic creditors. Some US$333 million (45 percent of GDP) of domestic debt was to be cancelled through land swaps. Land management companies—special purpose vehicles (SPVs) – were established specifically to manage the swaps which involved registering, valuing, and selling approximately 1,400 acres of land. The revenues from the land sales were to settle creditors' original claims. St. Kitts and Nevis debt relief was to be obtained primarily through the debt-for-land exchange. When finalized, the swap, along with the other completed components of the restructuring, was anticipated to result in debt-to-GDP declining from 160 percent to 100 percent.

Jamaica's debt exchanges, in 2010 and again in 2013, were both confined to the domestic portion of the debt. In both cases, only marketable domestic debt securities, excluding Treasury bills held by domestic residents and issued in Jamaica, were affected. The primary objective of both exchanges was to reduce the interest burden of the debt and to reduce the interest rate and exchange rate risk arising from large holdings of variable-rate and foreign currency denominated debt. The 2010 debt exchange also sought to smooth a severe bunching of payments in the following year.

Under the terms of the 2010 Jamaica Debt Exchange (JDX), 340 domestic securities were exchanged for 24 new "benchmark" bonds with extended maturities and lower interest rates (Government of Jamaica 2010). In 2013, under the terms of the National Debt Exchange (NDX), these bonds were subsequently exchanged for new benchmark bonds with even longer repayment periods and lower interest rates. Both the 2010 and 2013 debt exchanges had allocation rules that constrained bondholders to only exchange instruments of the same type. As a result, the exchange swapped existing local currency fixed-rate and variable-rate instruments as well as US dollar denominated and US dollar indexed bonds for new instruments of the same type. The rules' sole exception allowed holders of variable rate instruments to swap these into local currency fixed rate, variable rate, or inflation indexed bonds. The objective was to increase the holdings of fixed-rate debt and limit interest rate exposures as far as possible. An additional feature of the debt exchange was the removal of the call option on all fixed rate bonds to encourage bondholder participation.

Jamaica's 2010 debt exchange (JDX) enjoyed a voluntary participation rate of over 99 percent, making the exchange one of the most successful in the world. With no collective action clauses to push participation rates upward, the success of the exchange relied mainly on strong nationalism and a widespread view that a "national response was needed to address the country's pressing debt problem" (Jamaica Observer 2010). Jamaica had achieved similar success under its 2013 exchange with a 98.8 percent participation rate.

Outcomes

The outcomes of debt restructuring exercises have been varied across SIDS. However, an examination of key debt indicators shows that debt restructuring had limited impact on debt to GDP or on the debt service burden. Not surprisingly Seychelles which benefitted from debt cancellation had the most favorable indicators in the aftermath of its debt restructuring whereas most other SIDS, with the exception of Dominica (see table 9.3), saw their debt indicators worsen or remain virtually unchanged.

Table 9.3 Impact of debt restructuring on key debt indicators

	Year –2	Year –1	Year of Debt Restructuring	Year +1	Year +2
Antigua and Barbuda (2010)					
Public Debt-to-GDP	76.9	102.0	**90.6**	93.4	97.8
Interest payments to GDP	3.2	7.2	**2.1**	2.1	3.0
Dominica (2004)					
Public Debt-to-GDP	111.4	130.8	**116.0**	108.1	95.7
Interest payments to GDP	5.5	5.8	**5.8**	5.4	2.4
Debt Service to Revenues	n.a.	35.6	**35.5**	24.1	11.4
Grenada (2005)					
Public Debt-to-GDP	102.2	120.6	**110.3**	116.5	111.0
Interest payments to GDP	5.2	6.6	**2.0**	2.4	2.1
Debt Service to Revenues	27.1	40.4	**12.3**	13.3	14.1
Jamaica (2010)					
Public Debt-to-GDP	126.2	141.2	**143.0**	140.0	143.3
Interest payments to GDP	12.3	17.0	**11.1**	9.6	9.9
Debt Service to Revenues	111.3	134.7	**82.4**	85.9	67.3
Seychelles (2009)					
Public Debt-to-GDP	132.4	139.5	128.6	**82.5**	77.8
Interest payments to GDP	6.6	7.2	10.6	**6.1**	2.9

Note: n.a.: not available.

Source(s): International Monetary Fund (2014).

In terms of debt to GDP, only two SIDS saw significant declines in their debt to GDP. Seychelles debt declined from 128.6 percent of GDP in the year of the debt exchange to 82.5 percent two years after while Dominica's debt to GDP fell from 116.0 percent to 96 percent. Debt to GDP levels in Jamaica and Grenada remained virtually unchanged, while debt to GDP rose from 90.6 percent to 97.8 percent in Antigua and Barbuda.

Some SIDS, such as Grenada, saw their debt service burden reduce substantially. Grenada's interest payments fell by one-third to 2 percent of GDP and remained around that level in the ensuing two years. Debt service as a share of revenues fell to 12.3 percent, well below the pre-restructuring ratio of 40.4 percent in 2004.

Similarly, Seychelles' interest payments as a share of GDP fell from 11 percent at the time of restructuring to 3 percent in 2011. Jamaica's debt service burden also eased significantly in the years immediately after the debt exchange. Debt service as a share of revenues dropped to 82.4 percent in the year of the exchange from 134.7 percent a year earlier (2009). By 2012, the ratio had dropped to 67.3 percent or just half the value in 2010.

Conclusion

Although debt restructuring has been increasingly relied upon by SIDS to ease their debt burdens, in the main, such restructuring exercises have provided only temporary cash relief and little debt reduction. Only in Seychelles and, to a lesser extent, Dominica have debt to GDP levels fallen notably.

Jamaica's experience underscores the temporary nature of current restructuring operations, as its 2010 debt exchange was quickly followed by another exchange in 2013. Similarly, Grenada, which restructured in 2005, was again in debt distress in 2013, defaulting on an interest payment on a bond issued under its previous debt exchange. Grenada subsequently announced it would again comprehensively restructure its debt to achieve more substantial relief.[4]

SIDS, such as Dominica and Seychelles, that have opted for debt stock as well as flow reductions have generally had more durable declines in their public debt-to-GDP and fiscal burden ratios than market-access SIDS, such as Jamaica and Grenada, that chose only a flow reduction through rescheduling their debt (see table 9.4). Loss of market access appears to have been a significant factor for market-access SIDS in opting out of debt reduction (Robinson 2013). Also limiting the scope for debt reduction was the potential threat to financial sector stability. Concerns about financial sector stability were a major factor for SIDS undertaking domestic debt restructurings, such as Jamaica and Antigua

Table 9.4 SIDS—type of debt relief and impact on debt to GDP

Country	Type of Relief	Cut in Face Value (%)	Change in public debt-to-GDP (24 months after exchange) (%)	Public Debt to GDP 2012
Seychelles	Nominal stock reduction	50.0	-39.5-	82.5
St. Kitts and Nevis	Nominal stock reduction	31.8	n.a.*	89.3
Dominica	Nominal stock reduction	20.0	-17.5	74.8
Grenada	Flow reduction	0	0	109.5
Jamaica	Flow reduction	0	0	146.1
Antigua and Barbuda	Flow reduction	0	8.0	89.1

Note: *Two years have not yet elapsed since the implementation of the May 2012 debt exchange.

Source(s): IMF Regional Economic Outlook Update – Western Hemisphere (2014).

and Barbuda. Jamaica also cited constitutional constraints, namely a provision that prioritizes debt payments, as a reason for not pursuing a debt reduction strategy.

For flow-only SIDS, the gains from debt restructuring have quickly eroded and payments difficulties re-emerged. In contrast, Seychelles, following a haircut and lost market access, has seen more robust economic fundamentals following debt restructuring. Its debt has moved close to sustainable levels. A stronger basis now exists to re-enter the international capital markets and on favorable terms. The lesson may be that, for SIDS, debt reduction is the only viable option.

Notes

1. The Paris Club is an informal group of creditors from the major Western industrialized countries who provide a forum in which indebted countries can negotiate a restructuring or cancellation of their debts. Negotiations take place on a case-by-case basis.
2. The London Club is an informal group of private creditors, typically commercial banks, that join together to negotiate the restructuring of their claims against a sovereign borrower.
3. There is little information available as to why Caribbean SIDS did not negotiate Evian terms. One possibility is that the Paris Club may have viewed Caribbean SIDS debt payments difficulties as one of temporary illiquidity rather than as long-term insolvency, thereby making them ineligible for Evian terms. Alternatively, Caribbean SIDS may just not have sought to undertake the debt sustainability analysis necessary to negotiate Evian terms.
4. Belize, a small mainland Caribbean state, has also undertaken a second debt exchange within seven years, reflecting the immense difficulty of small states, in general, achieving debt sustainability.

References

African Development Bank. 2010. *Approves USD 10 Million for Credit Guarantee Programme for the Seychelles.* African Development Bank, February 17. http://www.afdb.org/en/news-and-events/article/approves-usd-10million-for-creditguarantee-programme-for-the-seychelles-6326/.

Government of Jamaica. 2010. "Jamaica Debt Exchange—Investor Presentation." *Ministry of Finance.* January 14. http://www.mof.gov.jm/sites/default/files/JDX_Investor_Presentation.pdf.

International Monetary Fund (IMF). 2004a. *Dominica: First Review under the Three-Year Arrangement Under the Poverty Reduction and Growth Facility and Financing Assurances Review Staff Report; Press Release; and Statement by the Executive Director.* IMF Country Report, Washington, DC: International Monetary Fund.

———. 2004b. *Progress Report to the International Monetary and Financial Committe on Crisis Resolution—April 2004*. Report to the IMFC Committee, Washington, DC: International Monetary Fund.

———. 2005. *Progress Report to the International Monetary and Financial Committee on Crisis Resolution—April 2005*. Report to IMFC, Washington, DC: International Monetary Fund.

———. 2012a. *St. Kitts and Nevis—First Review Under the Stand-By Arrangement and the Financing Assurances Review and Request for Waivers of Applicability and Modification of Performance Criterion—Staff Report and Press Release*. IMF Country Report No. 12/51, Washington, DC: International Monetary Fund.

———. 2012b. *St. Kitts and Nevis—Second Review under the Stand by Arrangement and the Financing Assurances Review, and Request for Waivers of Applicability—Staff Report and Press Release*. IMF Country Report No. 12/196, Washington, DC: International Monetary Fund.

———. 2013. *Regional Economic Outlook: Western Hemisphere—October 2013*. Washington, DC: International Monetary Fund.

———. 2014. *World Economic Outlook*. Washington, DC: International Monetary Fund.

Jamaica Observer. 2010. "Editorial—Support for the Debt Exchange." Jamaica Observer, January 13. http://www.jamaicaobserver.com/editorial/Editorial-for-Jan-13_7323967.

Paris Club. 2009. *Seychelles: Debt Treatment—April 16, 2009*. http://www.clubdeparis.org/sections/traitements/seychelles-20090416/viewLanguage/en.

———. 2010. *Antigua and Barbuda: Debt Treatment—September 16, 2010*. http://www.clubdeparis.org/sections/traitements/b-20100916/viewLanguage/en.

———. 2012. *Saint Kitts and Nevis: Debt Treatment—May 24, 2012*. http://www.clubdeparis.org/sections/traitements/saint-christophe-nieves/viewLanguage/en.

Robinson, Michele. 2013. "Assessment of Debt Restructuring Operations in Commonwealth Small States." In *Small States Digest*, by Commonwealth Secretariat, 10–21. London: Commonwealth Secretariat.

Chapter 10

Debt Sustainability and Sustainable Development in SIDS

Abdullahi Abdulkadri

Introduction

Increasing debt levels have a deleterious impact on the economies of nations, especially the developing countries, and it is arguably one of the most important factors that impede development and perpetuate poverty. In recognition of this, the international community has implemented a number of initiatives to help alleviate the debt burden of the most indebted poor countries with the expectation that debt relief will provide the macroeconomic environment conducive to a sustainable level of debt. However, poor countries are not the only ones with heavy debt burdens. Small Island Developing States (SIDS) are a group of countries with unique characteristics and common developmental challenges but different public debt experiences. Their uniqueness makes the attainment of sustainable development particularly challenging, although important, for the continued existence of most SIDS.

This chapter focuses on the interrelationship between debt sustainability and sustainable development in SIDS. Previous chapters have discussed the developmental impact of debt and the debt experiences of SIDS. An emerging issue from these chapters is the prospect for further growth in debt in SIDS. Amidst this expectation of growth in debt, what implications will high and growing levels of debt have on sustainable development in SIDS? In shedding light on the debt sustainability–sustainable development interaction, the chapter identifies the commonality of factors that impact sustainable development in SIDS and examines what role public debt has played in aggravating or ameliorating these factors and the consequential effects on sustainable development.

Following this introduction, a brief discussion of the concepts of debt sustainability and sustainable development is presented. That is followed by an elucidation of the characteristics of SIDS vis-à-vis the *three pillars of sustainable development* and public debt. The main thrust of the chapter then follows entailing a discussion of the interplay between debt sustainability and sustainable development. The chapter concludes with a summary of the main arguments in the chapter, including suggestions on how to utilize debt financing for sustainable development in SIDS.

Definition of Concepts

Debt sustainability has been described as the condition in which a borrower is able to service its debt "without exceptional financing or a major future correction in the balance of income and expenditure" (IMF and World Bank 2004, 8). Therefore, by definition, debt sustainability rules out any of the following situations: a required or anticipated debt restructuring; debt accumulation at a rate faster than the borrower's capacity to service it; and necessity to curtail wage bill and public expenditure through retrenchments in order to service debt obligations (even if nothing in the external environment changes). The forgoing suggests that the central theme of debt is solvency.[1] Simply put, debt sustainability is the ability of a country to meet its present and future debt obligations without requiring debt relief or accumulating arrears.[2] Hence, assessing debt sustainability requires projecting how debt dynamics are likely to evolve over time, thereby making debt sustainability a forward-looking concept that cannot be assessed with certainty (Wyplosz 2005).

Unlike debt sustainability with a fairly standard definition, the definition for sustainable development has evolved over time. Since the definition proffered by United Nation's World Commission on Environment and Development (WCED) in its 1987 report, *Our Common Future*, debt sustainability has been variously defined and these definitions scrutinized (e.g. Bugge and Watters 2003; Kates, Parris, and Leiserowitz 2005; Osofsky 2003; Taylor 1998; United States National Research Council 1999). In this chapter, the definition given by the 2002 World Summit on Sustainable Development, in its Johannesburg Declaration, is adopted. This definition introduced the concept of three pillars of sustainable development by advocating for "a collective responsibility to advance and strengthen the interdependent and mutually reinforcing pillars of sustainable development—economic development, social development and environmental protection—at local, national, regional and global levels" (United Nations 2002, 1). This definition advances the issue of sustainable development beyond the traditional focus on continuous economic growth that is environmentally sustainable to include issues of social development and environmental protection. The traditional view had relied on the notion that economic growth would produce a richer world that is ecologically stable (Hopwood, Mellor, and

O'Brien 2005). However, recent experiences of emerging economies have shown that economic growth is often accompanied by increased environmental degradation and widening social inequality.

The Developmental Challenges Faced by SIDS

SIDS constitute a unique case with respect to the challenges faced in attaining overall sustainable development and building resilience in their economies. These challenges and the underlying factors have been discussed in earlier chapters. This section highlights how the major factors impact on the three pillars of sustainable development and on public debt.

Economic Development

As a group, SIDS are vulnerable to economic shocks, thus placing them at a distinct disadvantage in comparison to larger and more developed countries. Their small sizes limit opportunities for economies of scale and their heavy reliance on tourism or export of primary goods forces undue specialization on these economies. These factors also foster a relatively high degree of dependence on international trade which further increases SIDS' susceptibility to global shocks. In 2007, SIDS on average derived 51 percent of their export earnings from international tourism receipts in comparison to other developing countries which earned less than 10 percent of the value of their exports from international tourism (United Nations 2010). With such a heavy dependence on one sector, a single global shock could have significant effect not only on the foreign exchange earnings of many SIDS but also on the domestic service industry that supports the tourism sector.

The location and small sizes of SIDS have more significant implications for the economies of these countries when the impacts of climate change are considered. An increase in the frequency or intensity of El Niño events would adversely affect agricultural production and food security, water resources, human health, and the coastal environment (UNFCCC 2005). Changing and unpredictable weather patterns and more frequent and more intense extreme weather events such as tropical storms, hurricanes and cyclones, storm surges, droughts, and floods represent major challenges for SIDS given the debilitating effects they have on livelihoods and infrastructure. Economic recovery after these events may take years and often translates into low output and high costs of reconstruction. For countries most affected, this presents a challenge for previous gains made in economic growth to be sustained. Several SIDS have faced these consequences[3] and with climate change their susceptibility to natural hazards is expected to increase.

Vanuatu and Tonga occupied the top two positions among countries with the highest probability of experiencing a natural disaster based on the World Risk Index (WRI).[4] Eleven other SIDS were among 35 countries categorized as having very high probability in 2011 (Birkman et al. 2011). By 2013, the number of SIDS among top 35 countries with very high probability had grown to 13 while Vanuatu and Tonga maintained their topmost rankings. These growing risks faced by SIDS have economic undertones as well. Over the three decades to 2009, several natural disasters caused enormous economic damage in SIDS and affected millions of people. Figure 10.1 shows the highest value of economic damage (equal or greater than 10 percent of GDP) caused by storms alone in SIDS during this period while figure 10.2 shows the highest disruption in livelihoods (equal to or greater than 10 percent of the population) caused by storms during the same period. Hurricane Gilbert caused the most costly destruction in St. Lucia in 1988 valued at 365 percent of the country's GDP. In terms of people affected, Cyclone Isaac caused the most disruption in Tonga in 1982 affecting more than 150 percent of the country's population; that is, every citizen was affected including nonresidents present in the country at the time of event. Despite advances in early warning systems, hurricanes and cyclones continue to pose major economic threats to SIDS. In 2004, Hurricane Ivan struck Grenada and destroyed 90 percent of the housing stock, 80 percent of agricultural assets, and left damage worth about US$900 million, or 150 percent of GDP. This was followed by a second hurricane ten months later. Prior

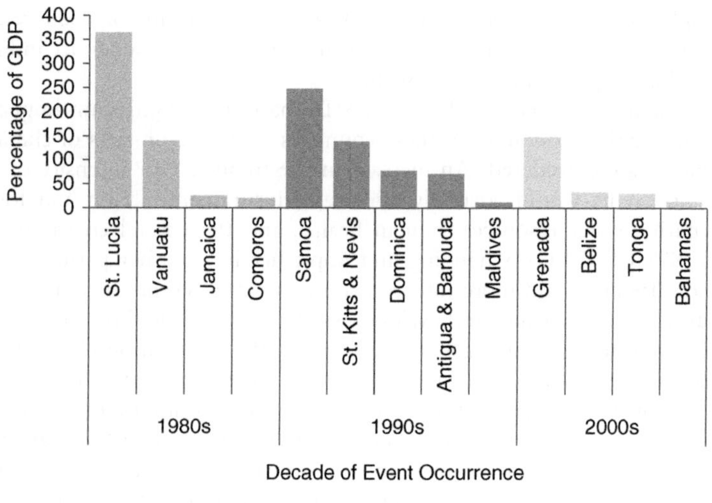

Figure 10.1 Damage caused by major storm events in SIDS (% of GDP)

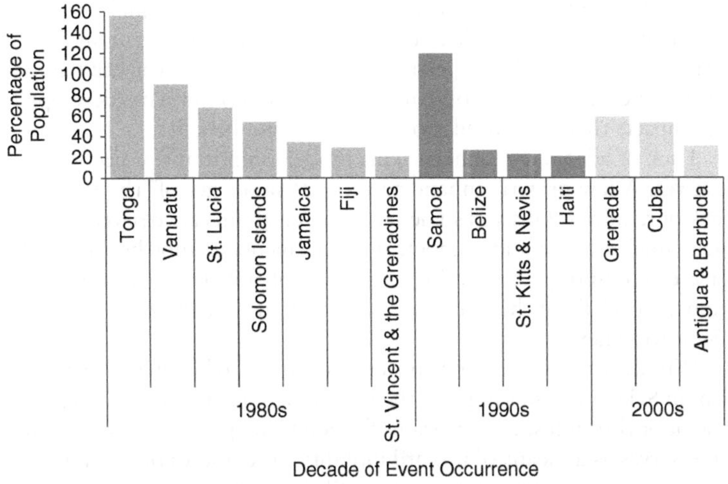

Figure 10.2 Persons affected by major storm events in SIDS (% of population)

to these events, Grenada was considered outside the Hurricane belt. It was among the five countries, out of 173, least likely to have a disaster, according to the 2013 World Risk Index. The Grenada example points to the fact that "low risk" does not constitute "no risk" and a singular event could have debilitating consequences for SIDS. When these events occur repeatedly, as has been the case for some SIDS, not only do they impede economic development, they make it unsustainable.

Evidence has shown that the negative impacts of climate change and natural disasters have compounded the existing sustainable development problems, worsened the poverty levels of many SIDS, and placed an increased burden on the already limited capacity of SIDS to build climate change resilience (UNDP 2010).

Social Development

With very few exceptions, SIDS often have limited human resources and institutional capacities to implement growth-enhancing projects that will lead to social development. As a result, SIDS are dependent on the international community not only for trade and official development assistance, but also for technological and technical assistance. Functional institutions are a requisite for social development and skilled manpower is required to achieve this. The situation in most SIDS is, however, that of limited technical skills aggravated by the migration of skilled professionals. The lack of employment opportunities in many SIDS (especially the LDCs) leads to a brain drain that deprives SIDS of qualified human resources, placing an undue burden on the education sector and forcing SIDS to rely on foreign expertise. Although migration has been noted to provide benefits in terms of remittances, the brain drain has developmental impacts that go beyond monetary costs and benefits.

Lack of adequate technical skills hinders important projects and initiatives. A few growth-enhancing projects never get off the ground for lack of required skills and some on-going programs are delayed when key personnel migrate and competent replacements cannot be found. This lack of continuity and the high turnover of staff in key technical positions, especially in the public service, make sustainable development difficult to achieve.

While not unique to SIDS, poverty continues to be a major social problem in SIDS necessitating that the government provides basic support for the poor through social programs funded by taxpayers. Although education serves as a means of upward mobility in socioeconomic status, children from poor households are more likely to drop out of school or have lower educational attainment, thereby perpetuating a cycle of poverty. Unfortunately, due to increased fiscal constraints, SIDS' governments have progressively withdrawn state subsidy for tertiary education making education and acquisition of high-level skills more expensive and beyond the reach of many poor households. In the face of high youth unemployment, the opportunity cost of education may appear too high for poor households and these households may be tempted to withdraw their children from school to engage in trade or employment. Not surprisingly, the United Nations (2010) indicated that out of 29 SIDS rated on the MDG of achieving universal primary education, six either regressed or made no progress.

Limited employment opportunities have nurtured social maladies including crime and violence in some SIDS with governments forced to devote scarce resources to national security and crime prevention in countries where these resources could be better used to promote economic growth that will create opportunities for poverty eradication and reduce income inequality. As these problem persist, the institutions set up to deal with them are stretched beyond capacity. This makes it difficult for the countries to achieve meaningful gains in social development and more likely to regress in their sustainable development efforts as social development needs compete with economic development goals and aspirations instead of promoting them.

Environmental Protection

Although SIDS occupy only 3 percent of the world's land area, they are known for their rich biodiversity, oceans, and coastal and marine resources. These are critical resources that can be harnessed for economic growth, yet there is always the contention between preservation of natural resources and exploitation for economic benefits. Tourism and agriculture, upon which a number of SIDS depend, are known for their detrimental environmental impacts and these sectors continue to exert pressure on the limited land resources of SIDS as more hotels and resorts are developed to cater for the tourism sector and more marginal lands are brought under cultivation with the aid of inorganic fertilizers and other chemicals. Faced with the prospect of heavy dependence on imported foods and the economic benefits from increased international tourist arrivals, protecting the environment from degradation and preserving its biodiversity often rank low on the priority of most governments. In circumstances where there is the political will to actively protect the environment, policy makers are often confronted with the dilemma of spending limited government revenue to preserve the ecosystem or exploiting this same ecosystem to generate the much needed revenue to fund the government and its other social programs. In most cases, the lure of revenue to be derived from the exploitation of natural resources tend to tip policy decisions in favor of rapid expansion of revenue generating opportunities that destroy the natural environment or reduce biodiversity as opposed to preservation that enhances environmental protection.

At the household level, environmental protection is generally of a low priority in times of economic hardship. Households in poverty tend to rely heavily on natural resources for their sustenance. With the relatively high cost of commercial energy, these households tend to use firewood and coal that are cheaper sources of energy for their heating and cooking needs. This contributes to deforestation and Greenhouse Gas emissions with consequences for the environment and the health of citizens.

The continued existence of a society depends heavily on a healthy environment and mass relocations of settlements in SIDS have been suggested as one way of dealing with the impeding effects of climate change. When human actions accelerate these climate change impacts or create new environmental problems, sustainable development suffers and SIDS, due to their peculiarities, have relatively fewer remedial options to fall back on. This fact, in particular, makes environmental protection critical for SIDS to attain sustainable development.

Public Debt

With the many challenges faced in attaining sustainable development—economic development, social development, and environmental protection—SIDS find themselves accumulating huge public debts. High exposure and vulnerability to global shocks worsen the fiscal position and the macroeconomic outlook of most SIDS. For example, the food, fuel, and financial crises during the 2007–2009 period reduced international trade, raised the cost of borrowing and resulted in significant increases in the price of food and energy. As net importers of food and energy (except Trinidad and Tobago), SIDS had to pay significantly higher prices for these commodities while earning less on tourism. Confronted with dwindling government revenues, rising expenditures and deteriorating foreign reserves, the governments of many SIDS sought credit from the domestic and international markets, sometimes at uncompetitive rates to shore up their fiscal balances.

Climate change, manifested in severe drought, extreme temperature, floods, and more intense and frequent storms, affects output and could lead to a decline of about 2 percent in real GDP per capita in low-income countries (ESCAP 2013). Reconstruction after major natural disasters has an effect on unplanned public expenditures that are often financed with overseas development aid (ODA) and loans. The frequency of these events sometimes means that the same infrastructure is impacted multiple times and requiring significant public expenditure to rebuild on each occasion. In areas of high vulnerability to the impacts of climate change, adaptation and mitigation options are implemented often at significant costs. Although the UN's Global Environment Facility (GEF) provides grants to eligible countries implementing these projects, the grants generally require cofinancing from the implementing country. For cash-strapped nations, this cofinance can only be provided through borrowing from other creditors.

High unemployment and emigration of skilled professionals reduce anticipated government revenue from taxes. Ineffective institutions for revenue collection also hinder the ability of SIDS to collect taxes due, accounting for a major leakage in government revenue. Revenue shortfalls are generally made up either through domestic short-term debts that tend to accumulate or tax increases that raise the cost of doing business and further lead to lower tax payments due to reduced economic activity.

Although public debt could result from a number of causes, debt accumulation beyond a sustainable level, on one hand, will adversely affect development and make development itself unsustainable. On the other hand, the quest to make economic development sustainable may drive debt accumulation up to a point where debt becomes unsustainable. Therefore, public debt plays a crucial role in the attainment of sustainable development.

Sustainable Development within the Context of National Debt

The harsh realities of environmental degradation, rising poverty, and social inequality, and the prevalence of increasingly fragile and vulnerable economies have brought sustainable development to the forefront of national and global discussion. Within this context, the role that national debt has played and continues to play in most nations' elusive quest for sustainable development has been highlighted. These concerns have resulted in concrete international efforts directed at addressing the debt burden of the world's poorest and heavily indebted countries through initiatives such as the Heavily Indebted Poor Countries (HIPC) Initiative (and subsequently enhanced HIPC) and the Multilateral Debt Relief Initiative (MDRI).

Prior to the introduction of HIPC, the most heavily indebted LDCs spent more on debt servicing than on health and education combined (IMF 2013a). Expectedly, this limited their ability to mobilize national resources for sustainable development. HIPC, MDRI, and other debt relief initiatives pursued by individual governments are already showing evidence of being effective in raising poverty reducing expenditures (PRE)[5] of governments (IMF 2013b). Figure 10.3 shows trends in the average debt service/GDP and PRE/GDP for HIPC over the period 2001–2012. In 2001, debt servicing represented 2.8 percent of GDP for HIPCs while PRE amounted to 6.5 percent of GDP. By 2012, debt service reduced to 1.3 percent of GDP while PRE/GDP had increased to 10.1 percent of GDP. Debt service/GDP steadily declined from 2001 reaching a minimum of 0.8 percent in 2010 but has been on an upward trend since 2010.

Comoros, Guinea-Bissau, Guyana, Haiti, and São Tomé and Príncipe are the only SIDS that have benefited from the HIPC Initiative.

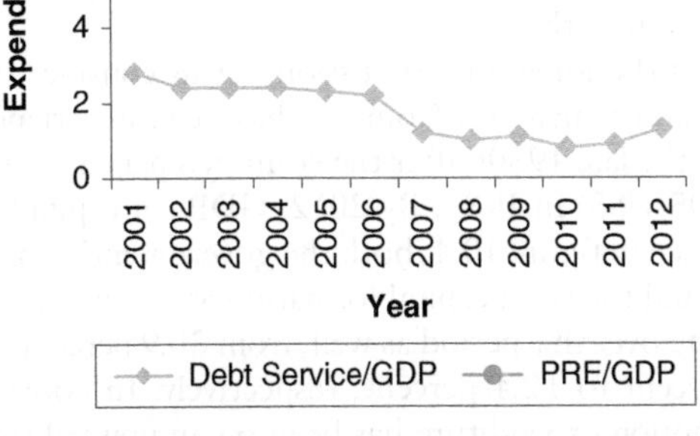

Figure 10.3 Trends in average debt service and poverty-reducing expenditure as a proportion of GDP for HIPCs (2001–2012)

Subsequent to the debt relief enjoyed under HIPC and MDRI, which created the budgetary leeway for increased public expenditure relative to debt-service payments, these SIDS have been able to increase their social and development expenditures. For example, Comoros reached HIPC decision point in 2010 at which time the country's debt service to GDP ratio was 1.0 percent and PRE/GDP was 6.6 percent. By the following year, Debt Service/GDP declined to 0.3 percent while PRE/GDP rose to 9.2 percent. Sao Tome and Principe reached HIPC completion point in 2007 with a Debt Service/GDP of 2.3 percent and PRE/GDP of 13.4 percent. In 2006, the country's Debt Service/GDP and PRE/GDP was 18.6 percent and 11.7 percent, respectively, but by 2012 its Debt Service/GDP had reduced to 2.3 percent while PRE/GDP increased to 20.6 percent (IMF 2013b). These are practical indications of the inherent conflict between debt servicing and public spending toward meeting the MDGs for SIDS with public debt burden. A high level of public debt constitutes a fiscal constraint for the government and limits public investment and expenditures on economic, social, and environmental programs that may facilitate sustainable development, address poverty, and help bridge income inequality (UNDP 2010). High levels of public debt are also usually associated with higher interest rates and could lead to lower economic growth with further implications for public revenues. This puts further strain on the government's budget and ability to service existing debt obligations and at the same time provide social services to the citizenry.

When countries have to choose between servicing their debt and expenditures on education, health, crime prevention, environmental protection and other social sectors, the challenges to economic development, social development, and environmental protection previously identified are either inadequately addressed or not at all addressed. Unfortunately, this hinders future economic growth trajectory as the underlying facilitators of economic development, namely; a skilled manpower, a safe and secure society, a healthy population and environment, and functional institutions, would not have been created or adequately developed to spur or sustain growth.

To buttress the foregoing discussion, let us consider the experiences of Comoros and Singapore. Comoros has been experiencing economic decline since the late 1980s after the country's per capita GDP peaked at $774 (in 2005 US$) in 1984. By 2012, GDP per capita had declined to $606. As shown in figure 10.4, both the government's final consumption expenditure and the fixed capital formation as a proportion of GDP have been declining over this period as well, from 30.9 percent to 15.3 percent and 28.5 percent to 12.4 percent, respectively. In contrast, household final consumption expenditure has been on an upward trend and growing more steeply since 2002 to reverse a downward trend in gross national

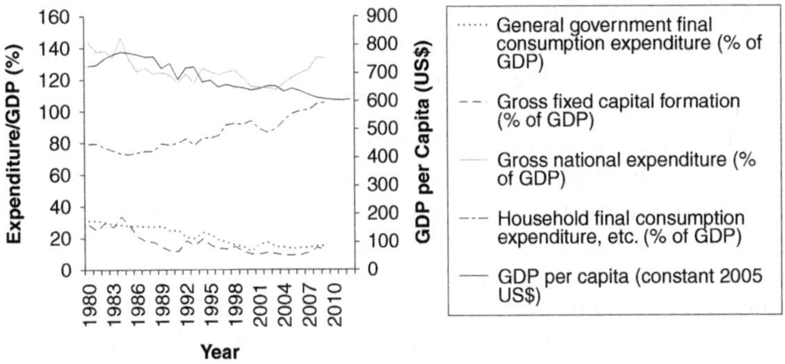

Figure 10.4 Trends in economic growth and national expenditures in Comoros (1980–2012)

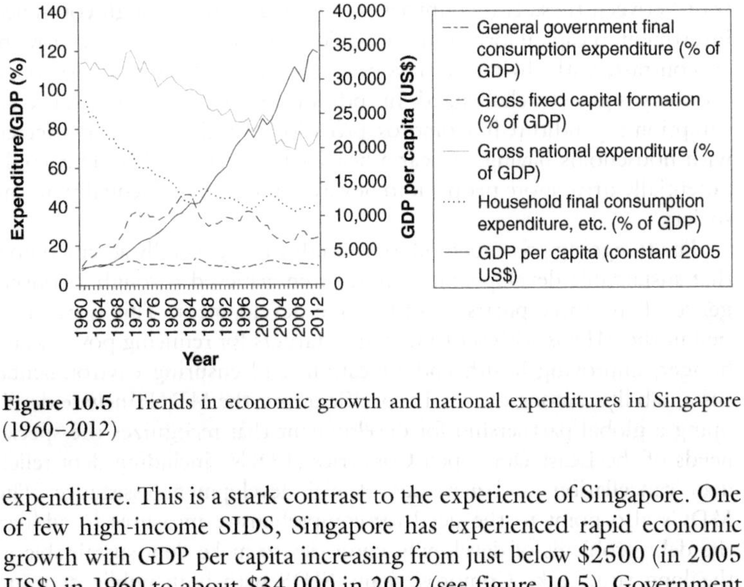

Figure 10.5 Trends in economic growth and national expenditures in Singapore (1960–2012)

expenditure. This is a stark contrast to the experience of Singapore. One of few high-income SIDS, Singapore has experienced rapid economic growth with GDP per capita increasing from just below $2500 (in 2005 US$) in 1960 to about $34,000 in 2012 (see figure 10.5). Government final consumption expenditure as a proportion of GDP did not evidence a secular trend, reaching a low of 8.2 percent in 1984 and a high of 13.5 percent in 1985. Gross fixed capital formation as a proportion of GDP, on the other hand, increased rapidly from 9.5 percent in 1960 reaching a high of 46.2 percent in 1984 before declining to 24.1 percent in 2012. With economic growth, household final consumption expenditure as a proportion of GDP declined rapidly from 94.8 percent to 42.3 percent between 1960 and 1983 and has been hovering in the low to mid-40s (percent) since then. Overall, gross national expenditure has steadily declined from above 100 percent of GDP between 1960 and 1983 to below 80 percent since 2003.

A comparison of the economic growth performance and public expenditures of Comoros and Singapore is therefore an instructive comparison. In Comoros, faced with a declining economy and a debt burden, the government has had to cut back on expenditures, both for consumption and capital investments. With a shrinking economy, households have had to shoulder a greater proportion of gross national expenditure which remained at 100 percent of the country's GDP since 1980. Conversely for Singapore, a booming economy with no debt burden,[6] the government's final consumption expenditure has kept pace with GDP while fixed capital formation was on an increasing trend for more than two decades and afterwards remained high. Among the reasons attributable to the different growth paths observed for Comoros and Singapore, the difference in capital investment over time stands out. A major implication of the contrast between these two countries is that Singapore, through continued investment in capital formation, was able to sustain its economic growth. In contrast, with dwindling investment in its capital stock, Comoros' economy is growth-challenged. In addition, as the government cuts consumption expenditure in Comoros, provision of social services is affected with households having to bear a heavier financial burden. This could potentially drive more people into poverty and increase inequality in the society.

The foregoing presents food for thought. It is generally accepted now that sustainable development is most often achieved through a convergence of the three pillars of sustainable development. This is exemplified in the MDGs with set time bound targets for reducing poverty and hunger, improving health and education, and ensuring environmental sustainability, among others. In specific terms, the MDGs include developing a global partnership for development that recognizes the special needs of the Least Developed Countries (LDCs), including debt relief, debt cancellation, and generous official development assistance. The MDGs also point to the need for comprehensive measures to address the debt problems of developing countries to make debt sustainable in the long run.[7] Like Comoros, many SIDS with high public debt are faced with fiscal constraints that force governments to make hard decisions between public spending and debt servicing. However, as discussed earlier, most SIDS have pressing development challenges requiring significant public expenditures which make debt servicing an enormous burden. The trade-off between debt servicing and public expenditure, in most cases, means that social support programs that would facilitate the achievement of MDGs are scaled back or cancelled outright. This constitutes a dilemma since sustainable development, and more specifically the attainment of MDGs, requires human and capital investment beyond that which most SIDS can provide in the absence of aid, grants, or loans. Moreover, the debt levels of most of these countries are already high enough such that more public debts are unsustainable. This is the precarious situation of many SIDS.

The United Nations has long acknowledged the uniqueness of SIDS and has taken meaningful actions to address the challenges they face in relation to the achievement of sustainable development. The attention paid to the growing debt levels of SIDS and the negative impact it has on their polity and development has, however, been less than adequate. Although 5 out of 36 HIPC beneficiaries are SIDS, a large number of SIDS continues to operate under heavy public debt burdens without being eligible for debt relief under the Initiative. While there are specific criteria (including income level and debt level) that define eligibility for HIPC, SIDS on average have comparable economic vulnerability with LDCs. Figure 10.6 shows the Economic Vulnerability Index (EVI)[8] of countries categorized as LDC, SIDS, and HIPC during the 1990–2011 period. The figure shows declining trends in economic vulnerability of the three groups over time. While LDCs and SIDS have similar average EVI values over the period, HIPC had consistently lower average EVI values than LDC and SIDS. However, while HIPC are predominantly LDCs, only a few SIDS are categorized as LDC. Even though SIDS are on average as economically vulnerable as LDCs, and more vulnerable than HIPC,

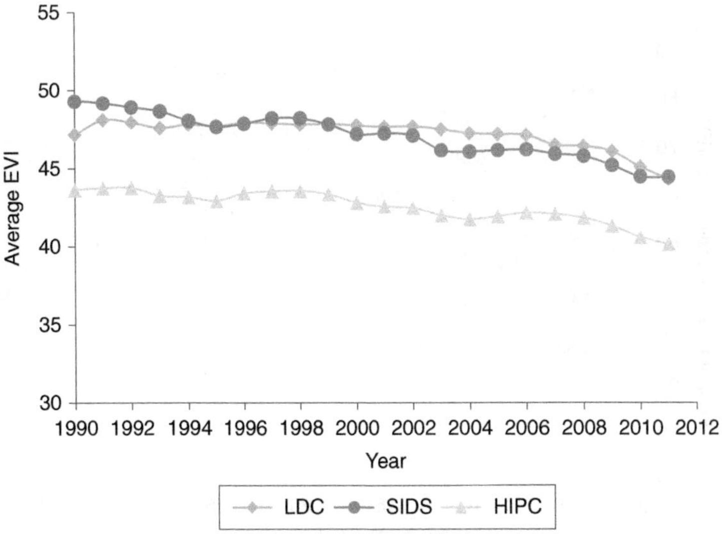

Figure 10.6 Trends in the economic vulnerability of groups of countries (1990–2011)

they are considered not poor enough to receive debt relief under HIPC. Meanwhile, SIDS continue to accumulate debt at unsustainable proportions. UNDP (2010) reported that 14 SIDS had debt to GDP ratios in excess of 60 percent with half of these countries having debt to GDP ratios above 100 percent. Particularly striking is the case of Guyana which, after benefiting from HIPC and reaching the completion point in 2003, still recorded a Debt to GDP ratio of 103 percent in 2010.

On average, SIDS have higher income levels than LDCs. Among 44 SIDS ranked by income level, only 3 (7 percent) were categorized as low-income countries against 12 (29 percent) categorized as high income. A further 15 (33 percent) were categorized as upper-middle income with the remaining 14 (31 percent) categorized as lower middle income (UN-OHRLLS 2013). This implies that more than half of SIDS are categorized as either upper-middle income or high-income countries. This relative high income level of SIDS belies their inability to bear the burden of public debt when economic vulnerability, natural disaster susceptibility, and other structural deficiencies and challenges are part of the context.

The cross-country relationship between economic vulnerability and susceptibility to natural disasters for SIDS is depicted in figure 10.7 using 2011 data on EVI and WRI for 26 countries. When the two outliers with the highest EVI (Kiribati) and WRI (Vanuatu) are excluded, the scatter plot shows a weak positive relationship between economic and natural disasters vulnerabilities. Moreover, income level seems to be

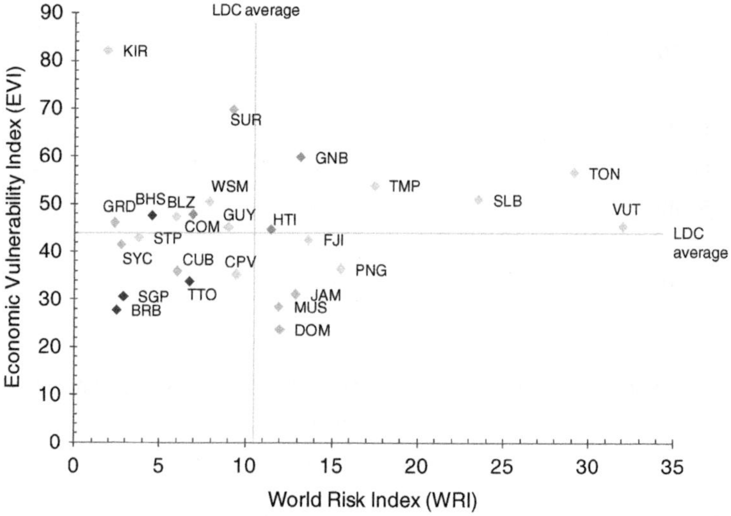

Figure 10.7 Relationship between income level of SIDS and country's EVI and WRI

a delineating factor. All the three low-income SIDS (Comoros, Guinea-Bissau, and Haiti) had economic vulnerability indices higher than the average of the LDCs. On the other hand, none of the high-income SIDS had a WRI higher than the average for the LDCs. The displayed data also suggest that either high income SIDS are able to minimize the risk faced from natural disasters (possibly due to their capacity to better cope and adapt) or that because they face low risk of natural disasters, their economies perform better. Unfortunately, most SIDS with high risk of natural disasters are lower middle income and upper-middle income countries that may be just one natural disaster away from an unsustainable level of public debt. Regardless, these countries will still be considered too rich to qualify for multilateral debt relief under HIPC.

Faced with fiscal constraints and increasingly unfavourable terms of multilateral external debts, many SIDS had turned to domestic and bilateral debts as sources. As their debt burdens became crippling, SIDS governments have resorted to restructuring of this portion of their debts. However, even with restructuring, the underlying causes of an unsustainable level of debt remained as governments battle to strike a balance between declining revenue and demand for public spending on social services. In such a situation, the prospect for many SIDS to achieve sustainable development is not promising. One way to assess SIDS' progress toward sustainable development is through the achievement of the MDGs. The latest summary on the progress of MDGs in SIDS indicates that SIDS are on-track in terms of gender, health, and certain education and environmental goals. Yet, as a group, SIDS have made less progress, and even sometimes regressed, compared to other groupings, especially in terms of economic growth, poverty reduction, and debt sustainability (United Nations 2010; United Nations 2013).

As we approach the year 2015, the target year for achieving MDGs, increased pressure has been exerted on SIDS to improve their performances. This comes at a time following a global recession in which most SIDS are in a worsened fiscal position and ultimately find themselves facing difficulties in financing their debt obligations while simultaneously financing a sustainable development agenda. This mismatch calls for new approaches to economic development and crisis and debt management. It also highlights the relationship among factors such as external shocks, internal structural deficiencies, and sociopolitical problems confronting the countries.

Conclusions

Climate change arguably is the single largest long-term threat to the existence of the human habitat and the economic success of countries, especially small islands. It has also become the de facto proxy for the implementation of sustainable development (United Nations 2010). The association between sustainable development and actions to address climate change can be mutually reinforcing (IPCC 2007) and the responses (mitigation and adaptation) to climate change, once appropriate, can help to concretely manifest sustainable development. Unfortunately, SIDS, while being highly vulnerable to natural disasters and significantly affected by climate change, lack the resources required to effectively adapt. At the same time, they are economically vulnerable to global shocks. These factors have helped to engineer unsustainable levels of public debts in most SIDS.

As they contend with high levels of public debt, SIDS also face challenges in their quest for sustainable development. Due to their unique characteristics, progress toward sustainable development has been slow in most SIDS. Functional institutions are a requisite for sustainable development and skilled human resources are required to achieve this. With high levels of migration of skilled professionals and high rates of unemployment, SIDS face a major challenge in relation to human capital development that have negative consequences for social development.

The lack of economic growth in countries with high public debt has given rise to the popularity of exploiting the natural resources for economic gains even if at the expense of the environment. While such policies may lead to economic growth, they tend to create environmental degradation and worsen income inequality.

Debt financing has been publicized by the international aid community as an important vehicle for mobilizing resources for public and private investment in developing countries; especially those with small domestic savings capacity and limited access to private capital.[9] Debt financing, therefore, offers an avenue for addressing the sustainable development–debt sustainability dichotomy in SIDS. However, given the current public debt situation of many SIDS, in order for debt financing to be successful, there has to be some level of debt relief or debt restructuring so that newly acquired debts can be sustainable and purposefully used for growth-enhancing investments. As HIPCs have shown, debt relief can also lead to increased poverty-reducing expenditures of governments which will serve as catalyst for sustainable development.

There is also the political element to debt sustainability. For debt and development to be sustainable in SIDS, there has to be a commitment at the highest level from governments, private sector stakeholders and the international funding agencies toward achieving this twin goal. This recognition is echoed in the eighth goal of the MDGs, which commits governments to: "Deal comprehensively with the debt problems of developing countries through national and international measures in order to make debt sustainable in the long-term."[10] More importantly, SIDS' governments must realize that there needs to be a strong balance between financing requirements for public expenditure and the need to preserve long-term debt sustainability. Hence, debt policies should include sustainable development measures against which public debts can be rated.

Two salient factors for debt sustainability are economic growth and debt relief. These factors are also essential for promoting sustainable development and redistributive growth in SIDS. Sustained economic growth is one of the most important ways in which governments are able to reduce public debt (Hileman 2012; UNDP 2010). However, the poor growth records of many SIDS (in comparison to other countries) have made "growing their way" out of indebtedness a herculean task. Additional policy measures are, therefore, required to support these countries to reduce their public debt burdens and foster sustainable development. It is in this spirit that the HIPC and MDRI were introduced, but so far only a few SIDS have benefitted, despite the challenges they face. A SIDS-specific debt relief initiative must be at the forefront of any policy agenda of the international aid community if debt sustainability, leading to sustainable development, is to be achieved in the near future.

Although SIDS are plagued by economic difficulties and confronted with barriers to development similar to those of developing countries, SIDS have their own unique vulnerabilities and characteristics which make sustainable development an illusion for many of them. An accommodation of these realities is essential in addressing sustainable development. Debt sustainability should be prioritized in any program designed for attaining sustainable development in SIDS.

Appendix 10.1: World Bank Country Codes

Country	Code
Antigua and Barbuda	ATG
Bahamas	BHS
Barbados	BRB
Belize	BLZ
Cabo Verde	CPV
Comoros	COM
Cuba	CUB
Dominica	DMA
Dominican Republic	DOM
Fiji	FJI
Grenada	GRD
Guinea-Bissau	GNB
Guyana	GUY
Haiti	HTI
Jamaica	JAM
Kiribati	KIR
Maldives	MDV

Country	Code
Mauritius	MUS
Papua New Guinea	PNG
St. Kitts and Nevis	KNA
St. Lucia	LCA
St. Vincent and the Grenadines	VCT
Samoa	WSM
Sao Tome and Principe	STP
Seychelles	SYC
Singapore	SPG
Solomon Islands	SLB
Suriname	SUR
Timor-Leste	TMP
Tonga	TON
Trinidad and Tobago	TTO
Tuvalu	TUV
Vanuatu	VUT

Notes

1. An entity is solvent if the present discounted value (PDV) of its current and future primary expenditure is no greater than the PDV of its current and future path of income, net of any initial indebtedness (IMF 2002).
2. See http://www.development-finance.org/en/topics-of-work/debt-strategy-information/debt-sustainability.html.
3. Samoa, St. Lucia, Grenada, Vanuatu, Tonga, and Maldives top the list of 180 countries with the highest economic losses on capital stock in relative terms due to natural disasters from 1970 to 2006 (Barittto 2008).
4. The World Risk Index is a measure developed by the United Nations University Institute for Environment and Human Security in Bonn, Germany. The measure indicates the probability that a country or region will be affected by a disaster and consists of four indicators of exposure to natural hazards, susceptibility, coping capacities and adaptive capacities to future natural events and climate change. See Birkmann et al. (2011).
5. PRE is defined differently for each country but generally consists of public spending on education, health and other social sectors that contribute to the reduction of poverty. The specific definition for each country is contained in the country's Poverty Reduction Strategy Paper.
6. Singapore does not have external public debt and its domestic public debt arises from it's deliberate decision to foster a debt market.
7. See http://www.unmillenniumproject.org/goals/gti.htm#goal1.
8. The EVI is a measure of the likelihood that a country's economic development would be hindered by external shocks. It is a composite of the size and likelihood of shocks, the exposure to shocks and the resilience to and capacity to withstand shocks. It is measured on a scale of 0 to 100, with 100 representing the highest level of economic vulnerability (Cariolle and Goujon 2013).

9. See http://www.un.org/esa/devaccount/projects/2004/0405K.html.
10. UNDP, MDGs, http://www.un.org/millenniumgoals/bkgd.shtml.

References

Barittto, F. 2008. "Disasters, Vulnerability and Resilience from a Macro-economic Perspective." *Background Paper for the 2009 ISDR Global Assessment Report on Disaster Risk Reduction.*

Birkmann, J., D. Krause, N. J. Setiadi, D. C. Suarez, T. Welle, J. Wolfertz, R. Dickerhof, P. Mucke, and K. Radtke. 2011. *World Risk Report 2011.* Berlin: Alliance Development Works.

Bugge, H. C. and L. Watters. 2003. "A Perspective on Sustainable Development after Johannesburg on the 15th Anniversary of Our Common Future: An Interview with Gro Harlem Brundtland." *Georgetown International Environmental Law Review* 15: 359–366.

Cariolle, J. and M. Goujon. 2013. "A Retrospective Economic Vulnerability Index, 1990–2011 Using the 2012 UN-CDP Definitions." *Working Paper 117.* Fondation Pour Les Etudes Et Recherches Sur Le Dévelopement International.

Economic and Social Commission for Asia and the Pacific (ESCAP). 2013. *Building Resilience to Natural Disasters and Major Economic Crises. Bangkok.* Thailand: United Nations Economic and Social Commission for Asia and the Pacific (ESCAP).

Hileman, G. 2012. "The Seven Mechanisms for Achieving Sovereign Debt Sustainability." *London School of Economics Working Paper, 162/12.*

Hopwood, B., M. Mellor, and G. O'Brien. 2005. "Sustainable Development: Mapping Different Approaches." *Sustainable Development* 13: 38–52.

Intergovernmental Panel on Climate Change (IPCC). 2007. *Climate Change 2007: Synthesis Report.* Geneva, Switzerland: Intergovernmental Panel on Climate Change (IPCC).

International Monetary Fund (IMF). 2002. *Assessing Sustainability.* Washington, DC: International Monetary Fund.

——. 2013a. *IMF Factsheet: Debt Relief Under the Heavily Indebted Poor Countries (HIPC) Initiative.* Washington, DC: The International Monetary Fund.

——. 2013b. *Heavily Indebted Poor Country (HIPC) Initiative and Multilateral Debt Relief Initiative (MDRI)—Statistical Update.* Washington, DC: The International Monetary Fund.

International Monetary Fund (IMF) and World Bank. 2004. *Debt Sustainability in Low-Income Countries—Proposal for an Operational Framework and Policy Implications,* February 3. Washington, DC: IMF and International Development Association (IDA).

Kates, R. W., T. M. Pariss, and A. A. Leiserowitz. 2005. "What is Sustainable Development? Goals, Indicators, Values and Practice." *Environment* 47: 9–21.

National Research Council. 1999. *Our Common Journey: A Transition Toward Sustainability.* Washington, DC: National Academy Press.

Osofsky, H. M. 2003. "Defining Sustainable Development After Earth Summit 2002." *Loyola L.A. International and Comparative Law Review* 26: 111–125.

Taylor, J. 1998. "Sustainable Development: A Model for China?" In *China in the New Millennium: Market Reforms and Social Development*, edited by J. A. Dorn. Washington, DC: CATO Institute.

United Nations. 2002. *Report of the World Summit on Sustainable Development, Johannesburg, South Africa, 26 August–4 September.* New York: United Nations Publication.

———. 2010. *Trends in Sustainable Development: Small Island Developing States (SIDS).* New York: United Nations.

———. 2013. *The Millennium Development Goals Report 2013.* New York: United Nations.

United Nations Development Programme (UNDP). 2010. "Achieving Debt Sustainability and the MDGs in Small Island Developing States." *UNDP Discussion Paper.*

United Nations Framework Convention on Climate Change (UNFCCC). 2005. *Climate Change, Small Island Developing States.* Bonn, Germany: Climate Change Secretariat.

UN-OHRLLS. 2013a. *Small Island Developing States (SIDS) Statistics.* United Nations Office of the High Representative for the Least Developed Countries.

———. 2013b. *World Development Indicators Database.* Washington, DC: World Bank. http://databank.worldbank.org/data/views/variableSelection/selectvariables.aspx?source=worldevelopment-indicators.

World Commission on Environment and Development. 1987. *Our Common Future.* Oxford: Oxford University Press.

Wyplosz, C. 2005. Fiscal Policy: Institutions versus Rules. *National Institute Economic Review* 191(1): 64–78.

Chapter 11

An Agenda for Debt Sustainability in SIDS*

Damien King and Michele Robinson

According to the theoretical and empirical literature, some amount of debt can be beneficial to developmental outcomes provided that the capital borrowed is productively employed, but high levels of debt tend to be deleterious to such outcomes (Tennant 2014). The now well-established debt overhang hypothesis posits that the growth-inducing benefits of low to moderate levels of external debt are eventually reversed as debt accumulates beyond a certain threshold (Deshpande 1997; Husain 1997). This threshold is theorized to occur when there is a debt overhang—the presence of debt sufficiently large that creditors do not expect with confidence to be fully repaid.

Many Small Island Developing States (SIDS) now have debts sufficiently large that they are experiencing debt overhang (Cabo Verde, the Seychelles, St. Kitts & Nevis, Grenada). This is not happenstance. Small island states seem to have a tendency to become indebted (Tennant 2014). The indebtedness is due primarily to two factors applicable in different parts of the world. In the AIMS and Pacific regions, thin domestic markets force governments to finance the development agenda with foreign borrowing, while in the Caribbean, the most highly indebted region in the world by far, poor management of contingent and non-central government liabilities forces the governments to absorb debt (King 2014).

This article proposes a comprehensive agenda of fiscal reform for SIDS designed to minimize the likelihood of accumulating unwarranted and unwanted levels of public debt. It is informed by the realization that the drivers of debt accumulation have their roots in the institutional structure of public decision making in small islands with weak governance capacity in a context of economic and natural vulnerability. The degree to which finance ministries have autonomy within the public sector and authority over other ministries will influence how they respond to the demands for expenditure and the ability to finance it. Their technical capacity will also have bearing on decisions about financing and debt management. An agenda for debt sustainability must therefore address the institutional foundations for fiscal outcomes and debt accumulation.

Default as a Strategy

Sovereign immunity allows the repudiation of debt with some degree of impunity since the assets of a sovereign cannot be attached to creditors. Highly indebted countries therefore often exercise the option to default or restructure their debts. Amongst small islands, there have been many such. Examples include Dominica (2004), Grenada (2005), Jamaica (2010, 2013), and Seychelles (2009).

Capital markets have historically been surprisingly forgiving of defaults/restructurings. Defaulters generally regain access to global capital markets after an average of three months and the interest premium for the affected countries is not prohibitive. Indeed, it is common for countries that restructure to receive a ratings upgrade, as was the case with Jamaica following its 2010 restructuring.

While restructuring is always an option, with or without the cooperation of the creditors, it is not a sustainable one. Despite the large number of island governments that have restructured, the debt problem persists. The reason for its persistence is that restructuring does not address the underlying causes of the propensity for indebtedness. More radical restructuring, involving haircuts, may reduce outstanding debt in the present, but it still leaves the country susceptible debt accumulation in the future.

Further, in practice, the restructuring often only delays payment obligations that eventually come due, and so is not a long-run solution. Meanwhile, the pool of debt that is amenable to restructuring may shrink in subsequent restructuring exercises, because the conditions of restructuring may restrict eligibility for future restructuring and also because creditors may become less sympathetic.

Moreover, the country suffers consequences from debt default/restructuring in other areas besides the cost and access to credit. Rose (2005) finds that trade declines following a restructuring by an average of 8 percent and the effect may last up to 15 years. This is perhaps due to the difficulty of getting trade credit following a restructuring. We have already referenced the risk to the financial sector from government default/restructuring, but Rose also points out that the effort to rebalance asset portfolios over the medium term impairs the performance of the financial sector. Finally, restructuring carries political risks, as evidenced by the frequency with which administrations that engage in debt restructuring lose power shortly after.

The conclusion is that the cycle of debt accumulation necessitating restructuring followed by renewed accumulation is not sustainable. At some point, the high-risk premium that is concomitant with this approach will be built into the interest rates at which island governments can borrow, and the long-run cost will be manifest even in the short run. Further, the economic and political consequences of actual and threatened restructurings will impair the prospects for long-run growth. Governments should, therefore, opt for better fiscal, liability, and debt management.

Fiscal Management

Creating fiscal space is critical to a program of debt reduction and to achieving debt sustainability. However, given the high demand for public goods and services by an often vocal and unforgiving electorate, political motivations frequently trounce sound financial management structures resulting in large fiscal imbalances and a buildup in public debt. A realignment of incentives could create consistency between the ideals of public sector financial management, the interests of a multidimensional electorate, and the electoral interests of politicians. Only popular support for sound financial management can drive political leadership that is consistent with these managerial ideals. Fiscal adjustment is likely to have short-term costs at the promise of benefits in the long term. There is therefore a clear time inconsistency problem where the electorate wants to see benefits now and politicians want to get elected. Enacting fiscal responsibility legislation that specifies performance targets and requires greater transparency and accountability in the government's fiscal operations is a means of creating an electoral counterbalance to the otherwise unlimited demands for seemingly free public services.

At present, few small island states have explicit institutional or legislative restrictions on fiscal policy. The overseas territories in the Caribbean, such as the Cayman Islands and Guadeloupe, have explicit expenditure guidelines and borrowing limits imposed by Britain and France respectively. As a consequence, the absence of fiscal autonomy has resulted in these islands having significantly lower debt levels than their peers amongst SIDS. Cyprus and Malta are bound by both the budget balance and debt rules of the Maastricht treaty. All the member states of the Eastern Caribbean Currency Union (ECCU) are supposed to adhere to the fiscal rules of their treaty. These group restrictions have been less effective. In addition, from time to time, some islands, such as the Seychelles and Jamaica, have come under the fiscal strictures of the International Monetary Fund as part of adjustment programs. The conditionalities embedded in those programs then provide boundaries to fiscal policy making.[2] Countries have a mixed record of adherence to the Fund's conditionalities.

Notwithstanding the above cases, most SIDS most of the time lack binding constraints on fiscal actions because of the absence of an explicit, legislated fiscal responsibility framework. Procedural, quantitative, and transparency requirements on fiscal policy are a means of cementing the basis for responsible fiscal policy making. Fiscal rules provide a counterbalance to short-sighted, opportunistic, and self-serving fiscal policy making.

An effective fiscal responsibility framework calls for more than just the legislated fiscal boundaries. Effective utilization of quantitative rules requires independent and technically equipped institutional oversight. Further, the effectiveness of transparency is dependent on public interest and support for reform, and a willingness and ability to read, interpret, and critically respond to an enhanced information flow.

Sanctions for violations have to be an effective disincentive to mismanagement. This means more than putting draconian fines and incarceration into law. The rules have to be clearly drawn to avoid ambiguity in interpretation or in the location of responsibility. And both the public and the prosecutor must hold responsible officers accountable for violations.

While the relevant laws suggest a demarcation of responsibility between public servants and the political directorate, any political encroachment on the administrative responsibilities of public servants makes it virtually impossible to locate responsibility, demand accountability and impose effective sanctions. The public service as in the Westminster ideal needs to be separated as far as possible from political influence. The capacity to hold public servants accountable is severely compromised if there is a blurring of the lines of responsibility between administrative and political leadership. This confusion of roles is likely where the political neutrality of the appointment process and the operational independence of the public servant within the law are compromised.

Consequences for departures from the fiscal rules need not be limited to sanctions. Budget sequestration, automatic expenditure cuts that kick in to prevent the incipient rules violations from being realized, are an effective means of keeping long-run fiscal targets on track. With the sequesters never being the most desirable outcomes, policy-makers are incentivized to avoid them.

In addition to establishing a fiscal responsibility framework to provide a long-term context and parameters for short-run fiscal policy making, accounting and procedural reform can improve fiscal outcomes. Systems for financial measurement and control need to be strong. Accrual accounting is an ideal. However, resource and institutional constraints imply, at best, a phased process of implementation. The recording and reporting of unpaid commitments on an ongoing basis offers significant enhancement at relatively low cost, even in the absence of a full accrual accounting framework.

Budgets should elevate the valuation of outputs (results) to the same status as inputs (financial resources used). A careful matching of inputs and outputs by way of corporate plans rising from the departmental level can be the basis for effective departmental "ownership" of budgets and for implementation consistent with resource availability and government's overarching strategic objectives. A contrary tendency to cut departmental budgets without reviewing the role, structure, and functions of Government tends to weaken ownership and undermine implementation.

Budgets for public bodies should be tabled at the time of tabling the government's budget, along with a corporate plan and an indication of projected financial results. These budgets should include estimates of projected dependence on central government resources.

Central Treasury Management will allow more efficient use of government's cash resources and reduce the need for, at least, some debt. Its implementation must, however, be balanced with the need to locate responsibility for financial mismanagement to ensure accountability.

Constitutionally independent institutions (answerable to Parliament) like Auditors-General and Contractors-General need to be supported, and are likely to require a greater infusion of financial, technical, and human resources for enhanced effectiveness. Stronger and more effective links with more efficient prosecutorial judicial systems are also required.

Liability Management

Fiscal responsibility legislation and the accounting and procedural reforms outlined above can effectively minimize the likelihood of debt accumulation due to fiscal slippage. But, as has been demonstrated, the bulk of the debt problems faced by most indebted SIDS do not arise from that source, but rather, from contingent liabilities that may originate outside of central government (King 2014). Moreover, even before contingent risks are actually realized, credit ratings agencies increasingly pay attention to them in determining the full risk of investing in a country's bonds, so these risks can raise the cost of debt. An agenda for sustainable debt therefore calls for better management of contingent liabilities.

The starting point for improved risk management in the public sector is to recognize and categorize the different types of contingent risks. One important distinction is that between inherent risks, such as arises when a natural disaster strikes and public expenditure is obligated, and acquired risks, exemplified by the issuance of explicit government guarantees or the propagation of implicit ones.

The first and easiest way to expunge risks that are not inherent in the business of governing is to divest assets and operations that carry such risks. Many contingent liabilities arise from the operations of public enterprises that are not even delivering public goods. This case is especially important for small islands with their limited resources for governing and therefore for managing an extensive of portfolio of public enterprises along with their contingent risks. Island governments should examine their list of public enterprises delivering private and non-essential services and divest those that carry excessive risk.

Where risks cannot be divested, they can sometimes be insured against. Once risks are identified and quantified, governments can seek explicit insurance, preferably internationally, against the particular risk. Where private insurers within the country may be impaired by precisely the event that causes the government to exercise its claim, countries may be able to insure each other.

To the extent that explicit insurance, such as that provided in the Caribbean by the Caribbean Catastrophe Risk Insurance Facility (CCRIF), is inadequate to sufficiently mitigate the fiscal risk of natural disasters, governments must themselves establish contingency funds in order to self-insure the uncovered risk. Such funds may either be explicitly set aside, which would then be invested in the capital market, or they may be virtually set aside, while the actual funds are available as part of the government's total treasury management.

Contingent liabilities must not only be identified, they must also explicitly be accounted for (either as an annual cost in the budget or on a net present value basis in the balance sheet) in public sector accounts. Government guarantees must then be granted only through parliament with full transparency. In that case, contingent expenditure would have to compete with other, budgeted expenditures for limited public resources. Both Barbados and Jamaica have already piloted limited initiatives in this direction.

Notwithstanding efforts to minimize risks by means of divestment, insurance, and provisions, governments may also recover a part of the cost of carrying these public risks. The guarantor, the government, could be compensated for bearing the risk either by means of an explicit payment (a premium), by requiring collateral, or by having a contractual claim on a share of the assets if the risk is not realized. Such a policy will not only compensate for the cost of the government being the insurer of last resort, but also reduce the demand for guarantees by filtering out unviable projects unwilling to pay. In cases where a government guarantee is a means of providing a desired subsidy, a better option is to charge for the guarantee and provide the subsidy separately so it can be subjected to the normal scrutiny of a public expenditure.

Better management of contingent liabilities should incorporate not only the improved practices outlined above, but also institutional reform that will facilitate the adoption of improved practices. The institutionalization ought to include centralized monitoring of contingent risks in an office with clear responsibility for such. That office should act on the basis of published guidelines for public enterprise borrowing and for the issuance of government guarantees. These guidelines will be based on the objective weighing of expected benefits with the level and cost of the risks. The decision process should be formal and transparent, preferably by a committee with membership outside the civil service based on professional qualifications. Finally, either the individual recommendations of the committee or the annual liability budget should require parliamentary approval.

Outsourcing a part of this responsibility would greatly improve outcomes. The governments should occasionally invite outside risk auditors to review both the risks themselves as well as the risk management processes of the centralized monitoring entity and publish its report.

Another element of a good institutional framework for managing contingent risks is a strong regulatory environment. Many contingent liabilities originate in industries that are serving a public purpose, such as utilities or finance. The primary responsibility for monitoring and controlling those entities lie with their statutory regulators. The legislative framework and technical capacity of the regulators should be brought up to be equal to the responsibilities.

Where to begin this process? Guidelines for better risk management can be gleaned from internationally accepted accounting standards for the public sector (IPSAS), which would strengthen the disclosure framework, as these standards require the identification, measurement, and disclosure of liabilities. The IMF's Code of Good Practices on Fiscal Transparency or the OECD's Best Practices for Budget Transparency also provide guidance.

Finally, a common source of contingent risk in any economy is bidirectional contagion between the financial sector and the fiscal accounts. This transmission channel is even stronger in small island economies with fragile financial sectors and weak public administrative capacity. The scope of the fallout from a crisis on either side, debt restructuring or a financial crisis, could be contained by de-linking the two spheres of activity and strengthening the respective institutions. Large financial institutions should limit their exposures to highly indebted governments. At the same time, financial regulators should ensure better provisioning and higher capital adequacy requirements in their respective financial sectors where the sector is heavily exposed in government securities.

Debt Management

While the suggested reforms to the management of both fiscal policy and off-budget contingent risks will help to avoid the accumulation of debt, improvements can be made to the management of existing debt. A comprehensive debt management strategy must accompany a government's stated debt management objective. Such a strategy should outline the manner in which the government intends to achieve its desired portfolio composition over the medium term based on its cost and risk preferences. Moreover, sound debt management practice requires governments to consider several alternative strategies to assess the potential variations in the cost of debt servicing based on various debt profiles.

Institutional reform would include that the relative roles of different public sector agencies and offices be delineated in order to avoid wasteful duplication. For example, while a central bank may act as an agent of the treasury in marketing and settling debt instruments, there is no logical basis for the central bank to duplicate record keeping, analysis, and articulation of debt policy. International best practice suggests a centralization of the core of debt management functions. This has often been within the treasury but a debt management unit may be established external to the Ministry of Finance. Such an approach separates debt management from fiscal policy, between which there are potential conflicts (e.g. with respect to cost versus risk minimization).

The institutional framework for debt management should assign clear responsibility for each of the important debt management functions.

1. Strategic planning, to determine the portfolio composition that conforms to the high-level objectives given concurrent and expected market conditions.
2. Risk management, to design a framework with responsibility for monitoring and managing risk exposures associated with exchange rates, interest rates, market access, and whatever events could jeopardize the achievement of the strategic plan.
3. Implementation, to determine and execute the borrowing plan that can match the articulated strategy mindful of the risk exposures.
4. Data management, to undertake the accurate recording, accounting, and publishing of all debt-related transactions and data.

The unit responsible for debt management is optimally organized along functional lines. The front office, guided by the agreed debt management strategy, is responsible for evaluating and negotiating new loans; the middle office takes care of analytical functions such as portfolio and risk analysis; and the back office executes loan servicing, accounting, and data gathering.

The disadvantages of small scale in the Caribbean may undermine the capacity to establish a centralized debt office with "firewalls" between functions. Where this precipitates a division of debt management functions between different institutions, there has to be a clear delineation of these functional responsibilities with unambiguous identification of the lead agency. This lead agency would preferably be located in the Ministry of Finance.

Effective management of public debt is associated with a wide range of public sector institutions, including the central bank (due to the fiscal costs of monetary policy and market operations) and the planning office (responsible for capital budget planning). Even where debt management is centralized, best practice still requires the establishment of an interagency committee for strategic oversight of debt management. This is especially important where there is fragmentation of the function between institutions, but is essential even in the case of a centralized debt management office.

The need for coordination is illustrated by potential conflicts between cost and risk minimization as fundamentally different objectives of debt management. Fiscal policy under pressure for short-term adjustments may prefer short tenor, external financing. This, however, exposes public debt to refinancing risk as well as foreign exchange risk later on. It is this conflict which informs the ideal of an independent debt management office and the imperative for an inter-agency strategic management committee.

While there is potential for conflict between fiscal and debt management, there are other areas in which there is consistency between their objectives. The fiscal objectives of the use of accrual accounting are entirely consistent with debt management objectives, namely, to avoid the disguised accumulation of public debt. Debt management would also be supportive of the avoidance of deferred financing, the careful management of public–private partnerships, the vigilant monitoring of public enterprises, and more effective treasury management systems. In general, short- to medium-term fiscal targets, legislated or otherwise, need to be consistent with medium- to long-run debt management objectives.

Debt management needs to be seen broadly in terms of a country's entire portfolio of public assets and not narrowly in terms of debt instruments only. Therefore, if there are public assets yielding a lower social and economic return than the highest cost debt, those assets should be privatized and the resources used to retire high cost debt. In a similar vein, if the returns to public capital expenditure at the margin exceed the interest cost of marginal debt, then surpluses should be dedicated to capital expenditure rather than paying down debt. These are the kinds of calculations that debt management with a sufficiently broad responsibility should consider. However, they depend critically on the specific mix of assets and opportunities in each country, so broad prescriptions should not be made.

Notwithstanding the importance of all the above, implementation of this debt management agenda will demand a high level of specialized technical and public policy expertise. Here, once again, we encounter and must acknowledge the limitation imposed by the scale of governance in small islands which militates against the sufficiency of such skills in each jurisdiction. However, the potential savings to be gained from optimal debt management at least provides the resources from which the capacity can be funded.

Conclusion

Many SIDS are severely constrained by a massive overhang of public debt. Bringing the debt to a sustainable level, one consistent with long-term economic growth and development, is a daunting task. Further, even if debt relief is provided, the risk of indebtedness will always threaten to derail the promise of growth given the economic and financial environment and the high vulnerability of island economies to external economic and natural shocks.

There is much that exposed and vulnerable governments can do. A more disciplined fiscal regime, more active debt management, careful management of contingent liabilities, and an enabling institutional framework are steps that small island governments can take to help manage, reduce, and prevent crippling debt.

Even more important than better policies and legislated rules, though, is the need to create a climate and an obligation of transparency and financial integrity in fiscal and debt management and public accounting. An environment of transparency provides an invisible cloak to constrain excess that can make legislated limits redundant. It should become both the obligation and practice of finance ministries to publish all information material to fiscal outcomes and the public debt. In so doing, the capital markets will automatically provide the implicit constraint of a swift response to unsound policies.

The objective of the institutional framework is to create a climate of autonomy, responsibility, and accountability. Public officials must be given the autonomy and authority to effect desired outcomes in their areas of responsibility. The framework should clearly identify where responsibility for each policy and outcome lies. While that responsibility can operate at different levels, to ensure accountability, it must converge at a single point. Once a unit or officer has both autonomy and responsibility, then he or she alone is accountable for the results. This arrangement would motivate better public policy-making.

Weak governance capacity due to the small size of the countries and scale efficiencies inherent in governing is a thread that weaves through much of the present discussion. The scale problem can be mitigated by cooperation among the islands, which would have the effect of exploiting economies of scale in management and smoothing the spread of best practices. Geographical proximity greatly facilitates such cooperation, however, so it is a realistic option only for the Pacific islands and even more for those in the Caribbean. Toward that end, there is some evidence of cooperation in Caribbean through regional institutions such as the Caribbean Development Bank, the ECCB Regional Government Securities Market, and the Caribbean Catastrophe Risk Insurance Facility. All are examples of regional cooperation that help Caribbean governments in the management of their debt.

But there is scope for more. Regional capital markets need to be deepened so as to expand debt management options. Placing debt in the capital markets of other islands will simultaneously support that objective while also minimizing the contagion risk to the local financial sector. Additionally, if governments established a regional institutional vehicle through which any expertise in debt management could be pooled, then the quality of debt management could be improved. Such an initiative could be pursued through an extension of the mandate of one of the existing regional bodies.

Fiscal management toward reduced debt loads should not distract governments from the ultimate goal of economic growth. GDP is the denominator in the debt/GDP ratio. Debt/GDP reduction can be achieved most easily through economic growth. So the efforts outlined here to manage debt must not sidetrack governments from the obligation to promote economic growth and development as a fundamental component of debt reduction. An additional 1 percentage point of annual GDP growth can take 20 percentage points off the debt/GDP ratio in ten years, roughly evenly divided between numerator reduction and denominator expansion.[3]

The indebtedness of many SIDS are ultimately due to policy, management, and institutional failures that magnify the impact by external and internal shocks. This essay has attempted to demonstrate that there are many ways in which the governments of these countries can institute reforms in policy frameworks, management practices, and institutional structures that can vastly improve the long-term outlook for the debt problem. All that remains is implementation.

Notes

* The material in this chapter draws heavily from one by the same authors and published as "Chapter 6: Solutions—An Agenda for Sustainability" in Caribbean Development Bank (2013).

1. For a discussion of the experience of debt restructuring in SIDS, see Robinson (2014).
2. As of this writing, mid 2014, Jamaica is the process of implementing a fiscal responsibility framework which has explicit rules for public debt and for loan guarantees. There are graduated benchmarks for these two variables which are scaled to GDP.
3. A higher GDP arithmetically reduces the debt/GDP ratio, but then stimulates virtuous debt dynamics in that less debt merits lower interest rates on government securities which therefore further reduces future debt.

References

Caribbean Development Bank. 2013. *Public Sector Debt in the Caribbean: An Agenda for Reduction and Sustainability.* Georgetown, Barbados: Caribbean Development Bank.

Deshpande, Ashwini. 1997. "The Debt Overhang and the Disincentive to Invest." *Journal of Development Economics* 52(1): 169–187.

Husain, Aasim M. 1997. "Domestic Taxes and the External Debt Laffer Curve." *Economica* 64(255): 519–525.

King, Damien. 2014. "Causes of Debt Accumulation in SIDS." In *Debt and Development in Small Island Developing States*, edited by Damien King and David Tennant. Basingstoke: Palgrave McMillan.

Robinson, Michele. 2014. "Does Debt Restructuring Work? An Assessment of Remedial Action in SIDS." In *Debt and Development in Small Island Developing States*, edited by Damien King and David Tennant. Basingstoke: Palgrave McMillan.

Rose, A. K. 2005. "One Reason Why Countries Pay Their Debts: Renegotiation and International Trade." *Journal of Development Economics* 77(1): 189–206.

Tennant, David. 2014. "The Debt-Development Dilemma: Challenges, Channels, and Complexities." In *Debt and Development in Small Island Developing States*, edited by Damien King and David Tennant. Basingstoke: Palgrave McMillan.

Chapter 12

Debt and Development in SIDS: Issues, Institutions, and Insights

Damien King

The claim often made by small island governments that they are naturally disadvantaged in the global economy appears not to have broad empirical support. Easterly and Kraay (2000) show that the growth rates of small states are no lower than, and controlling for other factors, their per capita GDPs are higher than, those of larger states. Rose (2006) concludes that small countries do not have systematically different outcomes for level of income, material well-being, health, education, and a number of other measures. While smallness appears not to create a disadvantage for standard of living in general, however, Small Island Developing States (SIDS) do show at least one economic malady that makes them stand out from their peers—relatively high levels of public debt.

While debt levels in the developing world have receded since the turn of the century, they remain stubbornly and unsustainably high in many SIDS. Sixty percent of the most indebted countries in the world are SIDS, whether you tally the top 5, 10, or 20. Fourteen SIDS have debt greater than 60 percent of GDP in 2014, the commonly accepted threshold of sustainability; seven are above 100 percent. SIDS, as a group, are among the most indebted class of countries in the world.

Notwithstanding the world's long experience with sovereign debt problems over the last half century, that of SIDS presents a peculiar difficulty because of the particular niche these countries occupy. SIDS are both more vulnerable and have fewer options. They endure higher vulnerability to external economic shocks because of their narrow production structures and to natural disasters due to their precarious geographies. Baritto (2008) points out that, of all the countries that have suffered economic losses from natural disasters over the previous 25 years, the top six (Samoa, St. Lucia, Grenada, Vanuatu, Tonga, and Maldives) are SIDS.

SIDS are mostly middle-income countries, however, and so do not have access to the institutions of debt relief that have been established for highly indebted poor countries. Moreover, the combination of being small and not abjectly poor ensures that the development spotlight never falls upon them. The SIDS debt problem is an important but largely silent one.

In order to begin to get a perspective on the cause of this problem, it is important to recognize that SIDS are a peculiar development grouping that encompasses both shared characteristics along with vast differences. They are defined by and therefore share small size[1] as well having a coastline. Size underlies many of the vulnerabilities they share: to external and internal economic shocks that are impactful due to the islands' necessarily narrow economies; and to sea level rise induced by climate change. An additional source of vulnerability is the dependence of many islands on tourism, a notoriously fickle industry. Since there are economies of scale inherent in public administration (Alesina and Wacziarg 1998), small countries must have either weak capacity or relatively large and costly public sectors, either of which can account for fiscal problems. A large number of the islands are of volcanic origin and/or lie in global storm paths and so are vulnerable there, too.

SIDS are not without advantages, though. Small size should promote social cohesion which reduces the conflicts that can motivate a misallocation of political attention and public resources. Smallness also creates an advantage in the ability of leaders to mobilize the population and resources toward problem-solving.

While SIDS share commonalities based on their size and island-ness, there are important differences. The Caribbean islands are close together geographically and culturally and are propitiously located near to the largest economy in the world in the United States. The Pacific islands are spread over a larger area and are remote both geographically and in terms of global shipping routes. The SIDS in the Atlantic, Indian Ocean, Mediterranean, and South China Sea (AIMS) are widely dispersed and diverse in structure and level of development.

A proper understanding of the debt problem in small island states will rest upon an appreciation of both their advantages and disadvantages as well as their commonalities and their differences. Numerous debt restructurings and fiscal consolidation attempts have failed to arrest the problem. This essay takes a comprehensive look at the problem in its peculiar context in order to understand its genesis and uncover solutions.[2]

The first section examines why debt is so damaging in small islands while the second describes the scale and nature of the debt problem. The third section investigates how and why these countries became so indebted so that the final section can propose an agenda of reform to address the problem.

Debt, Size, and Growth

At moderate levels, debt can be a constructive option in development-focused small island growth strategies. Even with their recent deepening over the last quarter of a century, domestic capital markets in SIDS still do not universally have the depth to accommodate the demand for development finance, especially in the tiny islands of the Pacific region. Additionally, external borrowing can facilitate public and private consumption smoothing in the face of disruptive external shocks—economic and natural (Bourne 2010). Domestic debt, like its external counterpart, can be growth-promoting up to a point. Singapore has used domestic securities to stimulate the development of a local capital market. Additionally, domestic debt provides some insulation from the vagaries of the international capital market.

Debt levels in many SIDS, however, are now much beyond the needs of development financing, consumption smoothing, and market making. This has put many island economies in a realm in which debt is constraining growth. While there remains some variation in the estimates of how much debt constitutes that threshold, there is widespread empirical support for the idea that beyond some level, debt begins to hold back growth (Kumar and Woo 2010; Patillo, Poirson, and Ricci 2002; Reinhart and Rogoff 2010).

One way this happens is through the "debt overhang" hypothesis (Deshpande 1997; Hussain 1997; Pattillo, Poirson, and Ricci 2004)—the idea that capital investment and growth-promoting reform are not worth the effort since part of the returns to them will have to be siphoned off to repay creditors. This keeps investment low and reforms unimplemented, spawning a high debt, low growth trap (Tennant 2014). Moreover, the debt overhang effect will be stronger in small economies since the likelihood that the claims are external will be greater the smaller the economy.

Too much debt impairs the prospects for growth in other ways. Capital expenditure is often the first category to be cut when resources need to be diverted to service debt (Fosu 2007). Maintenance of existing infrastructure can be scaled back nearly unnoticeably and new projects can easily be postponed. There is a saving in the present but at the cost of future growth.

The high cost of debt servicing also creates a precarious fiscal trajectory, engendering uncertainty. This arises, first, because of the risk of default, which raises the possibility of unavoidable, sudden fiscal contraction along with its attendant macroeconomic consequences. Uncertainty also arises from the risk of sudden stops—the withdrawal of international capital, which may result not only from domestic events but from contagion originating entirely outside of and therefore beyond the control of the indebted country. Consequences of this include capital flight and sharp currency depreciation (Tennant 2014).

High levels of domestic debt also threaten the solvency of the domestic financial sector, giving rise to the prospect of twin crises—public debt and private finance. This only serves to heighten the precariousness of the economy. Such uncertainty increases the riskiness of private investment and therefore shortens investment horizons. And as Patillo et al. (2004) have pointed out, it is the investment with the longer horizons that tend to have a higher payoff in terms of economic growth, so the resulting quick-return investment bias will be inimical to growth. Finally, to the extent that some of the debt is domestic, it crowds out domestic private investment. This is especially true in small states since the domestic capital market is small.

Even if high debt levels itself were not holding back the growth of small islands, the presence of debt may exacerbate the unsustainability of any particular growth trajectory. Small islands are particularly vulnerable to the effects of climate change. Mitigation, however, is costly and indebted governments are unlikely to prioritize that over servicing debt. Growth must also be socially sustainable (Abdulkadri 2014). This obliges governments to provide a social safety net and social services such as health and education. Retreat from the provision of these services falls disproportionately on the poor, worsening inequality and thereby threatening the social stability and political consensus upon which continued growth depends.

These effects, manifest while debt is being sustained, are usually exacerbated once the debt becomes unsustainable and debt contraction inevitable. Fiscal consolidation often impacts social services most severely (Tennant 2014). Given the higher relative cost of public administration in small states, the budget constraint will be tighter and the cost of expenditure diversion will be correspondingly greater. Even if, when the debt becomes unsustainable, the government succumbs to the temptation to inflate the debt away, rising inflation taxes the poor more heavily. So either way, the poor are disadvantaged, and growth becomes socially unsustainable.

The deleterious effects of high debt are not only quantitative, however. Managing debt and the threat of debt crises draw the energy of policy makers away from an agenda of economic growth, social equity, climate change, and human development. Severe indebtedness also reduces fiscal maneuverability and so inhibits the capacity of the government to respond to non-fiscal crises such as natural disasters and social disorder.

Beyond a threshold, then, debt constrains growth. But in small islands, the effect of debt on growth is likely to be much stronger. The downward pull on growth may not be immediate but by undermining the sustainability of growth in the long run, debt may exercise its most damaging effect of all.

The Profile of Debt

Debt/GDP ratios for 26 small island states, shown in figure 12.1, reveal both the extent of the debt problem as well as the variation by geographical region. Most of the islands, 9 of 13, in the top half of the table are in the Caribbean. All the Pacific islands are in the bottom half. Amongst SIDS, Caribbean countries are the most indebted and the Pacific islands are the least. The widely dispersed islands of the AIMS region are equally dispersed across the chart, giving it, as a region, average debt levels in

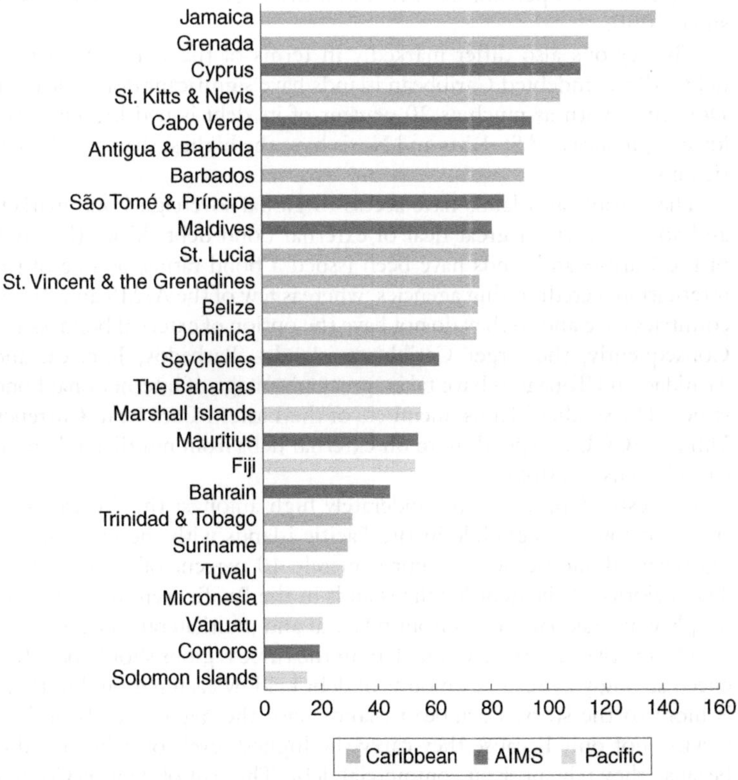

Figure 12.1 Debt/GDP, SIDS by region, 2013

between the other two regions, but also revealing that AIMS islands have the broadest range of debt outcomes.

If not for a debt restructuring that was promulgated by St. Kitts and Nevis in 2012, Caribbean islands would occupy all of the top three slots in the league table, with Jamaica and Grenada coming next. Caribbean debt was largely accumulated over the period surrounding the turn of the century and has remained high since then. The average debt of the Caribbean group in the table is 80 percent of GDP.

The most indebted countries in the AIMS region are Cyprus, Cabo Verde, and São Tomé and Principe. The wide range of indebtedness of the countries in the AIMS grouping reflects the wide geographical dispersion of the islands in this group and therefore the range of their circumstances. Average debt for the group is 69 percent of GDP.

The Pacific islands group had no country above the widely adopted 60 percent of GDP benchmark at the end of 2013. Marshall Islands, at 55 percent, is the most indebted. The average debt/GDP for the group is only 33 percent in 2013 and it has been on a downward trend since 2000.

The regions also differ markedly in terms of the structure of their debt. All the indebted Caribbean islands have significant domestic debt. Dominica, with as much as 20 percent of its debt issued locally, is the lowest. Jamaica and St. Kitts and Nevis have the highest shares of domestic debt.

The Caribbean islands have access to global sovereign bond markets and so can accrue a great deal of external bond debt. More than half of the Caribbean islands have been issued a bond rating by one of the international credit rating agencies, whereas few of the AIMS and Pacific countries have and so they do not have the option of external bond issues. Consequently, the larger Caribbean islands—Barbados, Jamaica, and Trinidad and Tobago—have taken great advantage of international bond issues. The smaller islands, members of the Eastern Caribbean Currency Union (ECCU), depend more on external debt from multilateral creditors (Robinson 2014a).

Domestic debt levels are moderately high amongst the AIMS countries and low or negligible in the Pacific islands with the exception of Fiji where domestic debt is approximately 40 percent of its total debt. The majority of the debt for the islands in the Pacific were borrowed on deeply concessionary terms from bilateral and multilateral sources.

The relative cost of servicing debt in the three regions should be influenced strongly by relative amount of debt held by each region, but there is more to the story. Caribbean islands have the highest levels of debt service, not only because they carry the highest levels of debt, but also because they rely most on commercial debt. The cost of debt service has exceeded the sustainable threshold of 15 percent of revenues in more than half of all Caribbean SIDS (Robinson 2014a).

AIMS SIDS had much lower debt service, mostly below the 15 percent threshold, with Seychelles and São Tomé and Principe being exceptions. Some countries in the AIMS region, in particular Cabo Verde and Comoros, were able to take advantage of concessional financing to lower their debt service (Barrett, Dawson, and McKenzie 2014).

Amongst the Pacific islands, the cost of servicing debt is even lower than their modest debt loads would explain because of the large amount of concessional bilateral and multilateral loans in their portfolios. The average for the region is around 10 percent of revenue and has been near that level for more than a decade (Prasad and Prakash 2014). Only Palau, the Marshall Islands, and Tuvalu have debt costs above 10 percent.

So many SIDS have executed debt restructurings, it may be considered one of the debt management strategies. There have been more than three dozen since 1990. Indeed, some countries have carried out more than one restructuring. (See Das et al, 2012, for a review of recent debt restructurings.) In the Caribbean, debt has been significantly restructured or relieved by Antigua and Barbuda, Dominica, Grenada (twice), Jamaica (twice), and St. Kitts and Nevis (Robinson 2014c). In the AIMS region, major restructurings or outright debt relief has benefitted Cabo Verde, Cyprus, Comoros (twice), and Seychelles. In the Pacific Islands, nearly all the islands have enjoyed debt forgiveness at some point, but only the Solomon Islands has carried out a formal debt restructuring.

Restructurings have been more frequent than debt relief because most SIDS do not qualify for forgiveness. The most indebted ones, largely in the Caribbean along with the Seychelles, have too little bilateral debt for Paris Club type relief to provide sufficient benefit. And most SIDS are too wealthy to qualify for HIPC. A debt exchange is the most popular tool of restructuring and, since 2000, has been utilized by Dominica (2004), Grenada (2005 and 2014), Seychelles (2010), St. Kitts and Nevis (2012), and Jamaica (2010 and 2013). Meanwhile, Antigua and Barbuda, Grenada, Seychelles, and St. Kitts and Nevis have used Paris Club for official bilateral relief but even then they did not qualify for the level of relief reserved for the poorest countries (Robinson 2014c).

Most restructurings have not included principal haircuts so debt/GDP would not initially be affected. Most of the restructurings are intended to deal with what has been perceived as a liquidity problem, not one of solvency. One reason for this is to exercise caution with respect to the possibility of contagion to the local financial sector which may hold a large portion of the domestic debt.

Whether under Paris Club, the Heavily Indebted Poor Countries (HIPC) initiative, the Multilateral Debt Relief Initiative (MDRI), or debt exchanges, and despite individual exceptions (such as Cabo Verde, Grenada, Solomon Islands, Seychelles), these arrangements have not resulted in sustained debt reductions for these islands as a group. For most islands, indicators remain largely unchanged or have worsened. Relief and restructurings, then, seem to provide only temporary relief from debt problems. This is supported by the observation that Grenada, Jamaica, and St. Kitts and Nevis have had to dip into the restructuring pool twice. The reason is clearly that a debt restructuring or debt forgiveness exercise by itself does nothing to address the factors that are driving debt accumulation in SIDS.

The Source of Debt

An examination of all the periods over the last 25 years where islands have experienced consecutive years of significant debt accumulation reveals a clear pattern of proximate debt drivers by region (King 2014). The primary cause of debt accumulation in the Caribbean has been existing debts assumed by the central government but originating outside of it—off-budget liabilities (OBLs). This factor is the largest contributor to debt in five of the six islands that have had the highest debt levels. Barbados is the only Caribbean island where the role of OBLs was exceeded by interest payments, and even there OBLs still accounted for a significant 24 percent of GDP worth of debt. Collectively, the need to finance primary deficits made no contribution to Caribbean debt, though the aggregation hides much variation. In particular, primary deficits were a contributor in Grenada and St. Kitts and Nevis (King 2014).

The outside liabilities arise from public enterprises, public–private partnerships, and from the private financial sector. In 2001, the Jamaican government absorbed the bad debts of its collapsed financial sector. In the early 2000s, the removal of European Union trade preferences for sugar and bananas saddled many island governments, including those in Jamaica, St. Vincent and the Grenadines, and St. Kitts and Nevis, with related losses. The 2009 failures of Antigua and Barbuda's Stanford Financial Group and Trinidad and Tobago based CL Financial affected nearly all the governments in Caribbean SIDS. Barbados along with the smaller islands of the Eastern Caribbean Currency Union were obligated to set aside liquidity for the 2009 failure of the British American Insurance Company.

In the indebted islands in both the AIMS and Pacific regions, with the single exception of Seychelles, the story is quite different. Seychelles' debt, which began to accumulate in the early and again in the late 1990s, was nearly all due to contingent liabilities, with the country having no primary imbalance to finance during those years. But for nearly all of the other islands in the two regions, financing primary deficits was the only significant contributor to debt. In AIMS, this consideration accounted for 62 percentage points of GDP of Cabo Verde debt, 50 for Cyprus, and 35 in the case of Malta. In the Pacific, the corresponding figures were 56 percentage points for Samoa and 40 for Solomon Islands (King 2014). A further disaggregation of the primary deficit in these countries reveals that capital expenditure accounted for most of the imbalance. Their more impressive growth rates over the past decade when compared with those in the Caribbean are likely related to their more productive use of debt.

In all the regions, a significant portion of the capital expenditure originates in the need for reconstruction following natural disasters. All the islands are prone to natural perils but those in the Caribbean and the Pacific even more so. In the Caribbean, Grenada, Jamaica, and St. Kitts and Nevis have suffered big losses from hurricanes in the past 15 years. In the Pacific, the Federated States of Micronesia, Samoa, and Tonga had to borrow heavily in 2000 for rehabilitation works after natural disasters. Altogether, natural disasters were responsible for adding an average of 24 percentage points of the debt/GDP ratio of small islands over the last three decades (UNDP 2010).

In summary, the data suggests that the AIMS and Pacific countries generally borrow to invest; the Caribbean and Pacific have to deal with capital reconstruction. And, in the Caribbean, governments absorb debt from contingent liabilities. This suggests that, apart from the largely concessional debt that some islands, particularly in the Pacific, receive for infrastructural and developmental purposes, much of the debt of the islands is not the result of intent or policy. Rather, it is often a residual, the unintended result of vulnerabilities and from institutional and administrative weaknesses.

Sustainable fiscal management requires, not only a degree of sophistication and capacity to identify and minimize contingent liabilities but also the ability to resist fiscal pressure coming from demanding, short-sighted electorates as well as those with legitimate but narrow interests. The institutions of governance do not force an alignment of principal (the wider public) and agent (fiscal management) (Caribbean Development Bank 2013).

One issue in at least some islands seems to be the concentration of power that derives from accomodating constitutional arrangements combined with weak civil administrations—a weakness that, as argued above, is inherent in small size. The facilitating constitutional arrangements would apply to most of the Westminster parliaments in the Caribbean. Under these circumstances the cabinet, and the Prime Minister in particular, wield enormous power. There are few checks in a hierarchical institutional arrangement and paternalistic political culture with little parliamentary oversight and inadequate civil service pushback.

As a consequence of these institutional structures, some island governments have the freedom to run up large debts without constraints and with little consequence. An examination of the pattern of indebtedness amongst small islands around the world supports the view that the institutions of public administration matter a great deal. Islands with effective constitutional constraints that derive from colonial relationships are the least indebted amongst SIDS—the Cayman Islands and the Turks and Caicos in the Caribbean, Bermuda in the Atlantic, and Guam in the Pacific Ocean are examples.

A culture of, as well as the legislative obligation for, transparency is another crucial institutional determinant of debt accumulation. Many of the off-budget liabilities that are dumped on central governments were not really contingent to begin with. When central governments tolerate loss-making public enterprises with implicit guarantees, the eventual absorption of the accumulated debts is not in fact contingent on anything other than the inevitable passage of time (Robinson 2014b). Deferred payments are another means of hiding debts under opaqueness. Sloppy accounting and weak accountability are the underpinnings of contingent debt.

Weak public administrative capacity is a recurring, facilitating theme in debt accumulation. Evidence comes from the observation that the relative size of government, as measured by the government expenditure share of GDP, is greater in small states. There are fixed costs in public administration. In the presence of fiscal constraints, this tempts governments to underpay civil servants and therefore results in a large but incapable civil service. There is little hope that such an administration can cope with the many contingent risks that lurk. It likely also produces the cost overruns on capital projects that contribute to debt growth.

Weak administration also allows debt, once accumulated, not to be managed in least cost or least risk ways, resulting in debt creating more debt, or at least, more than is strictly necessary. Debt grows because of risky debt structures and the absence of an articulated debt management strategy. In the absence of such, no one can be held accountable when debt becomes unsustainable. Measurements of debt management efficacy show that small states do lag behind that of even other developing countries (Prasad, Pollock, and Li 2013). Most SIDS governments, despite managing a great deal of debt, do not have a clearly articulated debt management strategy.

Debt accumulation may be driven by the need to fund development externally, by natural disasters, and by contingent liabilities. But it is facilitated, or even promoted, by both the inherent administrative weaknesses of size and hierarchical governance, and ultimately therefore by the failure to compensate for these with purposeful institutional change, fiscal constraints, and better management.

An Agenda for Sustainability

The debt problems in SIDS have been met with alternating cycles of debt restructuring and fiscal consolidation. We have already observed that numerous and variously designed restructurings, on the whole, have failed to solve the debt problem. Many SIDS have a tendency to become indebted and so restructurings that address liquidity issues are misguided and even those that tackle insolvency provide only temporary respite.

Meanwhile, the viability of the relief/restructuring strategy diminishes with each iteration since the conditions of new lending may restrict their eligibility for subsequent restructuring. Further, it imposes overt and hidden costs on a country. Ultimately, creditors will build the risk into the interest cost, raising the financial cost of this strategy. Repeated restructurings are costly in other ways, too. One way is that the level of trade usually falls, likely due to the increased difficulty of getting trade credit. Also, political parties that are responsible for restructurings tend to lose electoral support.

Fiscal consolidation at least addresses a source of debt increases, but it addresses the wrong one. Since some SIDS acquire debt purposefully for capital expenditure, either developmental or rebuilding, consolidation cannot and should not cut the capital portion of the budget. But the recurrent portion is not where the problem lies. Squeezing the recurrent budget more tightly is therefore extremely difficult and likely only to produce unsustainable cuts to public services or debilitating tax increases. Moreover, fiscal consolidation alone does not address the underlying cause of debt, leading to little debt reduction and ultimately to adjustment fatigue amongst the populace, undermining the appetite for reform entirely and so setting the country back.

Notwithstanding the constraints imposed by small domestic capital markets and the exigencies forced by vulnerabilities inherent in smallness, neither of which are themselves amenable to policy, there is much that small island governments can do to diminish the tendency to indebtedness. Improved fiscal management is a good starting point. Enacting a fiscal responsibility framework that sets a limit on the size of fiscal deficits and on the amount of public debt would provide a basis for resistance against the demands of an impatient electorate and the temptations to an unconstrained political administration. Key elements of such a framework would include the establishment of clear responsibility for violations between the political directorate and the administrative leadership and the provision for automatic corrections along with effective sanctions in cases of incipient and actual deviations (King and Robinson 2014).

Other procedural reforms can help to guard against fiscal slippage. Budgets across the public sector should be performance oriented, tracking outcomes as well as inputs (King and Robinson 2014). Central treasury management will allow for more efficient use of the public sector's liquid assets and reduce the need for some borrowing. Finally, accrual accounting will provide for a clearer picture of otherwise hidden, accumulating liabilities.

Contingent liabilities need to be better managed. This can be accomplished, first, by divestment of public enterprises that do not deliver public goods, thus eliminating the exposure to any contingencies inherent in their activities. Where responsibilities cannot be divested, risks that can be insured or hedged against should be. Finally, any remaining risks should be identified, quantified, and be the subject of explicit contracts with payments to the central treasury in exchange for the government guaranteeing, even implicitly, any liabilities.

There are two institutional reforms that will facilitate the adoption of improved risk management. First, a single entity must have full responsibility for monitoring and managing all contingent liabilities based on agreed, published guidelines, limited by an approved contingency budget, and using technical analysis to determine which risks have a positive expected value. Second, the regulatory bodies for public utilities should have the required technical capacity and be given the responsibility to monitor risks in their portfolio (King and Robinson 2014).

There is much room for improvement in debt management in nearly all of the indebted islands. It would begin with a political articulation of a clear debt management objective which would then inform the strategy to achieve it. Thereafter, a preferably centralized and independent debt management unit would be responsible for execution and management (King and Robinson 2014). Notwithstanding the need for that independence in order to avoid potential conflicts with the short run exigencies of fiscal administration, a coordinating inter-agency committee should ensure harmony between fiscal, monetary, and debt management.

Weak governance being a common thread, the island groups that are in closer proximity—the Pacific Islands and the Caribbean—should try to exploit opportunities for greater cooperation. Prime areas for such are in deepening their domestic capital markets, especially through cross placements and in the sharing of expertise and best practices across islands. Regional institutions to provide debt management services would also help.

Some islands have made a start at the required institutional reforms, though unevenly.[3] Jamaica introduced comprehensive fiscal rules in

2014 under the aegis of an IMF-supported programme. Progress toward improved debt management practices is more widespread. African SIDS have moved most aggressively toward improved debt management practices. Several Pacific islands, led by Tonga, have published debt management objectives. Jamaica, St. Kitts and Nevis, and Seychelles have improved the organization of their debt management units, while Barbados and Grenada have set up units for the first time. Jamaica, Mauritius, and Seychelles are far along in establishing a modern legal framework for debt management.

Ultimately, the surest way to sustain debt reductions is to exploit all opportunities for economic growth beyond debt reductions. Apart from the arithmetic reduction that comes from a higher valued denominator, virtuous debt dynamics kick in through lower interest rates.

Conclusion

Even though SIDS are not generally poorer than their larger peers, they do seem to have a tendency to end up with greater levels of debt relative to the size of their economies. This phenomenon has emerged mostly since 1990 with the vast expansion of the global capital market and the eagerness of even small economies to take advantage of it. At the same time, in many SIDS, their domestic capital markets have also expanded their capacity to absorb public debt.

The availability of cheap capital, however, does not by itself compel small island governments to borrow it more enthusiastically than their larger mainland counterparts. Nonetheless, island governments have indeed found cause to become the most indebted class of countries in the world. This outcome called for an explanation. This book has provided, and this chapter has summarized, a probable, multi-layered cause.

The explanation begins with the recognition of three key characteristics of small islands, the first of which is the small size of their domestic capital markets, notwithstanding their recent deepening in some cases. This forces governments to have to seek development capital externally. Another important characteristic is the extreme vulnerability of these small islands to both economic shocks and natural hazards. This creates a need for governments to spend repeatedly on social mitigation and capital reconstruction. The last characteristic is the scale economies inherent in public administration, which results in most SIDS having weak administrations unable to cope with contingent risks and efficiently manage large debt portfolios.

These characteristics interact to push these islands into debt to fund development and reconstruction and create a susceptibility to contingent liabilities. Fiscal consolidation is nearly doomed to be unsustainable under these circumstances. Institutional reform gets to the root of the problem and is a sustainable solution. Since one of the advantages of small size is the ease of mobilizing the citizenry, the islands should set about completing the required policy, institutional, and governance reforms. It is the only way for SIDS to have a hope of escaping the cycle of debt.

Notes

1. In this essay, our interest is in countries with a population of less than four million.
2. In doing so, it will draw upon the material in the other chapters in this volume.
3. This paragraph draws from Robinson (2014b).

References

Abdulkadri, Abdullahi. 2014. "Debt Sustainability and Sustainable Development in SIDS." In *Debt and Development in Small Island Developing States*, edited by Damien King and David Tennant. Basingstoke: Palgrave McMillan.

Alesina, Alberto and Romaine Wacziarg. 1998. "Openness, Country Size and Government." *Journal of Public Economics* 69(3): 305–321.

Barrett Kaymara, Altricia Dawson, and Sidonia McKenzie. 2014. "The Debt Experience of SIDS in the Atlantic, Indian Ocean, Mediterranean, and South China Seas." In *Debt and Development in Small Island Developing States*, edited by Damien King and David Tennant. Basingstoke: Palgrave McMillan.

Bourne, Compton. 2010. *A Discourse on Debt and Economic Growth in the Caribbean Community.* www.mona.uwi.edu/economics/notices/Presentations/Comptonpercent20Bourne.pdf.

Caribbean Development Bank. 2013. *Public Sector Debt in the Caribbean: An Agenda for Reduction and Sustainability.* Bridgetown, Barbados: Caribbean Development Bank.

Das, Udaibir S., Michael G. Papaioannou, and Christoph Trebesch. 2012, "Sovereign Debt Restructurings 1950–2010: Literature Survey, Data, and Stylized Facts." *IMF Working Paper* WP/12/203. Washington, DC.

Deshpande, Ashwini. 1997. "The Debt Overhang and the Disincentive to Invest." *Journal of Development Economics* 52 (1): 169–187.

Easterly, William and Art Kraay. 2000. "Small States, Small Problems? Income, Growth, and Volatility in Small States." *World Development* 28(11): 2013–2027.

Fosu, Augustin Kwasi. 2007. "The External Debt-Servicing Constraint and Public Expenditure Composition." *UNU-WIDER Research Paper No. 2007/36.*

Husain, Aasim M. 1997. "Domestic Taxes and the External Debt Laffer Curve." *Economica* 64(255): 519–525.

King, Damien. 2014. "Causes of Debt Accumulation in SIDS." In *Debt and Development in Small Island Developing States*, edited by Damien King and David Tennant. Basingstoke: Palgrave McMillan.

King, Damien and Michele Robinson. 2014. "An Agenda for Debt Sustainability in SIDS." In *Debt and Development in Small Island Developing States*, edited by Damien King and David Tennant. Basingstoke: Palgrave McMillan.

Kumar, Manmohan S. and Jaejoon Woo. 2010. "Public Debt and Growth." *IMF Working Paper WP/10/174.*

Pattillo, Catherine, Helene Poirson, and Luca Ricci. 2002. "External Debt and Growth?" *IMF Working Paper WP/02/69.*

Pattillo, Catherine, Helene Poirson, and Luca Ricci. 2004. "What are the Channels through which External Debt Affects Growth?" *IMF Working Paper WP/04/15.*

Prasad, Abha, Malvina Pollock, and Ying Li. 2013. Small States—Performance in Public Debt Management. *Policy Research Working Paper 6356*, Washington, DC: World Bank.

Prasad, Biman and Kushneel Prakash. 2014. "The Debt Experience of SIDS in the Pacific." In *Debt and Development in Small Island Developing States*, edited by Damien King and David Tennant. Basingstoke: Palgrave McMillan.

Reinhart, Carmen M. and Kenneth S. Rogoff. 2010. "Growth in a Time of Debt." *NBER Working Paper Series 15639.*

Robinson, Michele. 2014a. "The Debt Experience of SIDS in the Caribbean." In *Debt and Development in Small Island Developing States*, edited by Damien King and David Tennant. Basingstoke: Palgrave McMillan.

———. 2014b. "Institutional Underpinnings of Debt in SIDS." In *Debt and Development in Small Island Developing States*, edited by Damien King and David Tennant. Basingstoke: Palgrave McMillan.

———. 2014c. "Does Debt Restructuring Work? An Assessment of Remedial Action in SIDS." In *Debt and Development in Small Island Developing States*, edited by Damien King and David Tennant. Basingstoke: Palgrave McMillan.

Rose, Andrew K. 2006. "Size Really Doesn't Matter: In Search of a National Scale Effect." *Journal of the Japanese and International Economies* 20(4): 482–507.

Tennant David. 2014. "The Debt-Development Dilemma: Channels, Challenges, and Complexities." In *Debt and Development in Small Island Developing States*, edited by Damien King and David Tennant. Basingstoke: Palgrave McMillan.

UNDP. 2010. *Achieving Debt Sustainability and the MDGs in Small Island Developing States.* New York: United Nations Development Programme.

Contributors

Editors

Damien King is a senior lecturer and Head of the Department of Economics at the University of the West Indies, Mona, where he teaches graduate and undergraduate courses in macroeconomics and International Trade and Finance. Damien is also the director of the Caribbean Policy Research Institute, a think tank that focuses on public policy issues affecting the Caribbean region. His research has been in the areas of debt, poverty and distribution, and computable general equilibrium modeling and he has published in the *Journal of Development Studies*, *World Development*, and *Social and Economic Studies*, and also in edited collections.

David F. Tennant is a senior lecturer in Economics at the University of the West Indies, Mona Campus, and was previously an economist in the Jamaican Ministry of Finance and Planning. He has a strong background in quantitative and qualitative research methodologies. His areas of specialization include: Financing Development in Emerging Markets; Financial Crises; Micro, Small and Medium Enterprise (MSME) Development; and Pensions Reform. He has published papers in journals such as: *Journal of Banking and Finance; Journal of Economics and Business*; *Journal of International Financial Markets, Institutions and Money; World Development*; *Applied Economics*; *Applied Financial Economics*; *Journal of Developing Areas; The Empirical Economics Letters; Journal of Economic Issues;* and *Social and Economic Studies.*

Authors

Abdullahi Abdulkadri is a senior lecturer in the Department of Economics at The University of the West Indies, Mona Campus. He specializes in applied economics and carries a diverse research portfolio covering topics in agricultural economics, energy economics, environmental economics, health economics, and climate change. His studies in these areas have informed national and regional programs and policies. Widely published and internationally renowned, Dr. Abdulkadri received the Distinguished Past President Award from the Southwestern Economics Association for Outstanding and Continuous Service to the Organization in 2013.

Altricia Dawson is an economic analyst in the area of fiscal management. Though specializing in fiscal management, she has a diverse research portfolio covering areas in behavioral operations, family owned business (FOBs) in the Caribbean and the open data initiative. Additional research interests include development economics, industrial organization, macroeconomics, and operations management. She has also worked in the field of operations management focusing on logistics and inventory management. Ms. Dawson holds an MSc in Economics and a BSc in Operations Management from the University of the West Indies.

Biman Prasad is former Professor of Economics and Dean of the Faculty of Business and Economics at the University of the South Pacific. Professor Prasad holds adjunct Professorial positions at the James Cook University and Griffith University in Australia and Punjabi University in India. Professor Prasad has published extensively on economic and social issues on the Pacific Islands.

Kario-Paul Brown, a finance professional with experience in econometric modeling, economic forecasting, investment strategy formulation, and asset pricing. He has worked with the Caribbean Development Bank (CDB) and the Jamaica Money Market Brokers (JMMB) Group. He holds an MSc in Economics specializing in financial economics and international finance from the University of the West Indies and a BSc in Economics and Statistics also from the University of the West Indies. His research interests include financial modelling, debt simulation analysis, financial asset pricing, and the use of derivatives in risk management.

Kaymara Barrett holds a BA in History and Archaeology and an MSc in International Economics and Law, both from the University of the West Indies. She is currently employed at the Latin American Caribbean Centre (LACC) where her work focuses on improving grass roots relations between Jamaica and Venezuela. Her research interests include Development Economics, Agricultural Economics and International Trade.

Kushneel Prakash is a teaching assistant in the School of Economics at the University of the South Pacific. He obtained his Bachelor of Commerce degree and is currently pursuing his Master's Degree in

Economics from the same university. His research interests are in the areas of economic modeling, applied econometrics, development economics, and international trade.

Michele Robinson is an independent consultant in public debt management. She has worked extensively in the area of public debt management providing technical assistance to a wide range of countries in Africa, the Caribbean, the Pacific and Eastern Europe. Prior to becoming a consultant, she was the first head of the Debt Management Unit at the Bank of Jamaica, subsequently working at the Commonwealth Secretariat and then as debt management adviser at the Ministry of Finance, Jamaica. She has written numerous technical reports, papers, and articles in the area of public debt management, more recently on the public debt of small states. She is a graduate of the University of the West Indies in Jamaica and Queens University in Canada.

Sidonia McKenzie is currently an economist in the Monetary Analysis and Programming Department at the Bank of Jamaica (BOJ). Prior to joining BOJ, she served as a research assistant in the Department of Economics at the University of the West Indies, Mona Campus where she conducted research in environmental economics and climate change. Her research interests include exploring economic development issues in the Caribbean and other Small Island Developing States and applied econometrics with macroeconomic applications.

Index

Note: b, n, t respectively denote box, note, and table.

9781137397126